AF389169

Wellness, Fitness, Body Composition, Training and Performance Monitoring to Improve Athletes Life Quality

Wellness, Fitness, Body Composition, Training and Performance Monitoring to Improve Athletes Life Quality

Editors

Rafael Oliveira
João Paulo Brito

Basel • Beijing • Wuhan • Barcelona • Belgrade • Novi Sad • Cluj • Manchester

Editors
Rafael Oliveira
Sport Sciences School
of Rio Maior
Polytechnic Institute of Santarem
Rio Maior
Portugal

João Paulo Brito
Sport Sciences School
of Rio Maior
Polytechnic Institute of Santarem
Rio Maior
Portugal

Editorial Office
MDPI
St. Alban-Anlage 66
4052 Basel, Switzerland

This is a reprint of articles from the Special Issue published online in the open access journal *International Journal of Environmental Research and Public Health* (ISSN 1660-4601) (available at: www.mdpi.com/journal/ijerph/special_issues/Athletes_Life).

For citation purposes, cite each article independently as indicated on the article page online and as indicated below:

Lastname, A.A.; Lastname, B.B. Article Title. *Journal Name* **Year**, *Volume Number*, Page Range.

ISBN 978-3-0365-9395-1 (Hbk)
ISBN 978-3-0365-9394-4 (PDF)
doi.org/10.3390/books978-3-0365-9394-4

Contents

About the Editors . . . . **vii**

Preface . . . . **ix**

Rafael Oliveira, Ruben Francisco, Renato Fernandes, Alexandre Martins, Hadi Nobari and Filipe Manuel Clemente et al.
In-Season Body Composition Effects in Professional Women Soccer Players
Reprinted from: *Int. J. Environ. Res. Public Health* **2021**, *18*, 12023, doi:10.3390/ijerph182212023 . **1**

Renato Fernandes, João Paulo Brito, Luiz H. Palucci Vieira, Alexandre Duarte Martins, Filipe Manuel Clemente and Hadi Nobari et al.
In-Season Internal Load and Wellness Variations in Professional Women Soccer Players: Comparisons between Playing Positions and Status
Reprinted from: *Int. J. Environ. Res. Public Health* **2021**, *18*, 12817, doi:10.3390/ijerph182312817 . **11**

Ronaldo Kobal, Rodrigo Aquino, Leonardo Carvalho, Adriano Serra, Rafaela Sander and Natan Gomes et al.
Does the Number of Substitutions Used during the Matches Affect the Recovery Status and the Physical and Technical Performance of Elite Women's Soccer?
Reprinted from: *Int. J. Environ. Res. Public Health* **2022**, *19*, 11541, doi:10.3390/ijerph191811541 . **24**

José E. Teixeira, Luís Branquinho, Ricardo Ferraz, Miguel Leal, António J. Silva and Tiago M. Barbosa et al.
Weekly Training Load across a Standard Microcycle in a Sub-Elite Youth Football Academy: A Comparison between Starters and Non-Starters
Reprinted from: *Int. J. Environ. Res. Public Health* **2022**, *19*, 11611, doi:10.3390/ijerph191811611 . **33**

Ryland Morgans, Patrick Orme, Eduard Bezuglov, Rocco Di Michele and Alexandre Moreira
The Immunological and Hormonal Responses to Competitive Match-Play in Elite Soccer Players
Reprinted from: *Int. J. Environ. Res. Public Health* **2022**, *19*, 11784, doi:10.3390/ijerph191811784 . **44**

Igor Junio Oliveira Custódio, Renan Dos Santos, Rafael de Oliveira Ildefonso, André Andrade, Rodrigo Diniz and Gustavo Peixoto et al.
Effect of Small-Sided Games with and without the Offside Rule on Young Soccer Players: Reliability of Physiological Demands
Reprinted from: *Int. J. Environ. Res. Public Health* **2022**, *19*, 10544, doi:10.3390/ijerph191710544 . **57**

Ruby T. A. Otter, Anna C. Bakker, Stephan van der Zwaard, Tynke Toering, Jos F. A. Goudsmit and Inge K. Stoter et al.
Perceived Training of Junior Speed Skaters versus the Coach's Intention: Does a Mismatch Relate to Perceived Stress and Recovery?
Reprinted from: *Int. J. Environ. Res. Public Health* **2022**, *19*, 11221, doi:10.3390/ijerph191811221 . **67**

Yi-Hung Liao, Chih-Kai Hsu, Chen-Chan Wei, Tsung-Chieh Yang, Yu-Chi Kuo and Li-Chen Lee et al.
Monitoring the Changing Patterns in Perceived Learning Effort, Stress, and Sleep Quality during the Sports Training Period in Elite Collegiate Triathletes: A Preliminary Research
Reprinted from: *Int. J. Environ. Res. Public Health* **2022**, *19*, 4899, doi:10.3390/ijerph19084899 . . . **79**

Danilo A. Massini, Anderson G. Macedo, Tiago A. F. Almeida, Mário C. Espada, Fernando J. Santos and Eliane A. Castro et al.
Single- and Multi-Joint Maximum Weight Lifting Relationship to Free-Fat Mass in Different Exercises for Upper- and Lower-Limbs in Well-Trained Male Young Adults
Reprinted from: *Int. J. Environ. Res. Public Health* **2022**, *19*, 4020, doi:10.3390/ijerph19074020 . . . **94**

Giada Ballarin, Luca Scalfi, Fabiana Monfrecola, Paola Alicante, Alessandro Bianco and Maurizio Marra et al.
Body Composition and Bioelectrical-Impedance-Analysis- Derived Raw Variables in Pole Dancers
Reprinted from: *Int. J. Environ. Res. Public Health* **2021**, *18*, 12638, doi:10.3390/ijerph182312638 . **104**

About the Editors

Rafael Oliveira

Rafael Franco Soares Oliveira is an adjunct professor at Escola Superior de Desporto de Rio Maior, Instituto Politécnico de Santarém (Sport Sciences School of Rio Maior, Polytechnic Institute of Santarem), an integrated member of the Life Quality Research Centre, and a collaborator of the Research Centre in Sport Sciences, Health Sciences, and Human Development. He earned his PhD in Sports Sciences at the University of Beira Interior, Covilhã, Portugal. His research activity has been focused on exercise physiology, testing, control, and prescription; load monitoring and quantification; strength and conditioning; and sports training methodology. Currently, he is the author of more than 100 published scientific articles (indexed in the *Journal of Citation Reports or Scientific Journal Ranking*) and about 20 books or book chapters.

Furthermore, in 2022 and 2023, he was awarded first place in research and development by the Polytechnic Institute of Santarem. He was also awarded an honorable mention as the researcher with the best quotation in scientific production in 2022 and first place as the researcher with the best quotation in scientific production in 2023 by the Life Quality Research Centre.

João Paulo Brito

João Paulo Moreira de Brito is a full professor at Escola Superior de Desporto de Rio Maior, Instituto Politécnico de Santarém (Sport Sciences School of Rio Maior, Polytechnic Institute of Santarem), an integrated member of the Life Quality Research Centre, and a collaborator of the Research Centre in Sport Sciences, Health Sciences, and Human Development. He earned his PhD in Sports Sciences at the University of Lleida, Lleida, Spain.

His research activity has been focused on the following: exercise physiology; physiological assessment of athletes from different sports; determination of the physical and physiological profiles of elite athletes; subjective and objective load monitoring and quantification; and strength training to improve performance in male and female athletes. Currently, he is the author of more than 80 published scientific articles in indexed journals and about 17 books or book chapters. Furthermore, he received several awards for achieving a good quotation in scientific production in recent years.

Preface

Research on training and competition quantification, well-being/wellness, fitness, and body composition can be found in the literature, especially in soccer with male athletes. However, there are many other sports that intend to produce knowledge on these topics that also deserve merit. Moreover, more studies should include women athletes rather than only men.

Although some studies have already been produced, the present Special Issue prioritizes research focusing on the understanding of how training and match/competition monitoring can help improve athletes' quality of life. Therefore, the inclusion of well-being/wellness, health, fitness, and body composition variables in the studies was recommended, as well as the analysis of relationships between well-being/wellness (e.g., sleep quality, stress, muscle soreness, fatigue, and mood), exercise training programs, and usual training/match external and internal measures such as total distance, distances at different threshold speeds, accelerometry-based variables (e.g., acceleration, deceleration, and player load), session-rated perceived exertion, heart rate, and others.

Therefore, the aim of this Special Issue, which now constitutes a reprint, was to compile and provide new and updated knowledge on wellness, fitness, body composition, training, and performance monitoring and how they can be used to improve athletes' quality of life.

We believe that this reprint provides relevant information for several sports, namely, soccer, speed skating, triathlon, strength training, and pole dancing, in order to apply better strategies to the training process, helping athletes to improve or maintain their quality of life.

Rafael Oliveira and João Paulo Brito
Editors

International Journal of
Environmental Research and Public Health

MDPI

Article

In-Season Body Composition Effects in Professional Women Soccer Players

Rafael Oliveira [1,2,3,*], Ruben Francisco [4], Renato Fernandes [1,2,5], Alexandre Martins [1,2], Hadi Nobari [6,7,8,9], Filipe Manuel Clemente [10,11] and João Paulo Brito [1,2,3]

1   Sports Science School of Rio Maior–Polytechnic Institute of Santarém, 2040-413 Rio Maior, Portugal; rfernandes@esdrm.ipsantarem.pt (R.F.); alexandremartins@esdrm.ipsantarem.pt (A.M.); jbrito@esdrm.ipsantarem.pt (J.P.B.)
2   Life Quality Research Centre, 2040-413 Rio Maior, Portugal
3   Research Center in Sport Sciences, Health Sciences and Human Development, 5001-801 Vila Real, Portugal
4   Exercise and Health Laboratory, CIPER, Faculdade Motricidade Humana, Universidade de Lisboa, 1499-002 Lisbon, Portugal; ruben92francisco@gmail.com
5   University of Trás-os-Montes e Alto Douro, 5001-801 Vila Real, Portugal
6   Department of Exercise Physiology, Faculty of Educational Sciences and Psychology, University of Mohaghegh Ardabili, Ardabil 56199-11367, Iran; hadi.nobari1@gmail.com
7   HEME Research Group, Faculty of Sport Sciences, University of Extremadura, 10003 Cáceres, Spain
8   Sports Scientist, Sepahan Football Club, Isfahan 81887-78473, Iran
9   Department of Physical Education and Sports, University of Granada, 18010 Granada, Spain
10 Escola Superior Desporto e Lazer, Instituto Politécnico de Viana do Castelo, Rua Escola Industrial e Comercial de Nun'Álvares, 4900-347 Viana do Castelo, Portugal; filipe.clemente5@gmail.com
11 Instituto de Telecomunicações, Delegação da Covilhã, 1049-001 Lisboa, Portugal
*   Correspondence: rafaeloliveira@esdrm.ipsantarem.pt

check for **updates**

Citation: Oliveira, R.; Francisco, R.; Fernandes, R.; Martins, A.; Nobari, H.; Clemente, F.M.; Brito, J.P. In-Season Body Composition Effects in Professional Women Soccer Players. *Int. J. Environ. Res. Public Health* **2021**, *18*, 12023. https://doi.org/10.3390/ijerph182212023

Academic Editor: Paul B. Tchounwou

Received: 25 October 2021
Accepted: 15 November 2021
Published: 16 November 2021

Publisher's Note: MDPI stays neutral with regard to jurisdictional claims in published maps and institutional affiliations.

Abstract: This study aimed to analyze anthropometric and body composition effects in professional soccer women players across the early and mid-competitive 2019/20 season. Seventeen players (age, height, body mass, and body mass index of 22.7 ± 6.3 years, 167.5 ± 5.6 cm, 60.7 ± 6.6 kg and 21.6 ± 0.2 kg/m^2) from a Portuguese BPI League team participated in this study. The participants completed $\geq 80\%$ of 57 training sessions and 13 matches. They were assessed at three points (before the start of the season (A1), after two months (A2), and after four months (A3)) using the following variables: body fat mass (BFM), soft lean mass (SLM), fat-free mass (FFM), intracellular water (ICW), extracellular water (ECW), total body water (TBW), and phase angle (PhA, 50 Khz), through InBody S10. Nutritional intake was determined through a questionnaire. Repeated measures ANCOVA and effect sizes (ES) were used with $p < 0.05$. The main results occurred between A1 and A2 for BFM (-21.7%, ES = 1.58), SLM (3.7%, ES = 1.24), FFM (4%, ES = 1.34), ICW (4.2%, ES = 1.41), TBW (3.7%, ES = 1.04). Furthermore, there were significant results between A1 and A3 for FFM (4.8%, ES = 1.51), ICW (5%, ES = 1.68), and PhA (10.4%, ES = 6.64). The results showed that the water parameters improved over time, which led to healthy hydration statuses. The training load structure provided sufficient stimulus for appropriate physical fitness development, without causing negative disturbances in the water compartments.

Keywords: phase angle; female; body fat mass; fat-free mass; intracellular water; rated perceived exertion

1. Introduction

Soccer is considered one of the most popular sports worldwide [1]. To improve soccer athletes' performance and health, the assessment of anthropometric and body composition variables have been considered crucial [2]. Especially at a competitive level, body composition is an important component in an athlete's fitness and health profile and in each sport, performance is improved in specific ways in order to prevent injury risk [3].

Special attention has been paid to body fat mass (BFM) and fat-free mass (FFM). It is well known that an increased fat mass compromises performance, while increased muscle mass can promote the development of strength and power, which are important for players' performance [4–6]. According to a recent consensus statement, there are no values for BFM or FFM that should be followed, even more if we consider female soccer players [6]. For instance, in female player from US collegiate division 1, BFM of 16% was observed. In fact, the consensus statement added that it is not known what kind of body composition changes during the season may impact positively or negatively on the performance of the players [6].

Moreover, the interest in assessing other body composition variables, such as total body water (TBW), intracellular water (ICW), and extracellular water (ECW), to monitor hydration status in athletes has grown. For example, some studies have shown that ICW is a good predictor of strength and power in athletes [7–9].

Thus, considering the importance of body composition for athletes, frequent assessments should take place. This will allow coaches and athletes to know the development of body composition throughout the sports season and adjust training programs to prevent injuries and enhance sports performance.

Over the last decades, women's participation in sports has greatly increased. Although scientific research on women soccer athletes is growing, it is still limited [5,10–13]. Coaches and sports-related professionals should be aware of gender-specific questions and needs for optimizing performance. Especially at an elite level, few data have been used to show changes in anthropometric and body composition in women soccer players during the in-season [14]. To the knowledge of the authors, if the variables mentioned above and the training load variables, such as rated perceived exertion (RPE), were considered simultaneously, no studies were found. According to a recent report, performance measured by training and/or match data and body composition assessment could help soccer coaches and their staff to provide proper information for each player [6].

Specifically, internal load, which is one of the two dimensions of load monitoring (the other is external load), is a crucial psychophysiological part of the training load monitoring processes. One of the most frequently used variables to access internal load is RPE or the session-RPE (s-RPE, multiplication of RPE by session duration). This measure is a valid, reliable, and sensitive approach to quantify and qualify the internal load while using a simple questionnaire [15].

Knowledge of the essential characteristics for successful women's team soccer performance is useful to coaches, physicians, nutritionists, and exercise physiologists to improve their knowledge about women soccer athletes.

Therefore, this study aims to analyze the variations on anthropometric and body composition variables and their relationship with internal load in elite women soccer players across early and mid-competitive in-season using bioelectrical impedance analysis (BIA).

2. Materials and Methods

2.1. Experimental Approach to the Problem

This was an analytical and observational cohort study. The training sessions were performed during a five-month period, from September to January (early-to-mid-season) due to the COVID-19 pandemic, which provoked the disruption of training sessions and matches and the suspension of the season in March. The anthropometric and body composition assessments were conducted on three different occasions: the first week of September (before the start of the season, A1), after two months (the second week of November, A2), and two months after A2 (the third week of January, A3). All the assessments were performed under the same room and environmental conditions (place, time of day, order of tests application, temperature, and relative humidity, respectively, 22–24 °C and 55–65%) and by the same examiner. The players did not perform any other complementary training sessions during the period analyzed.

2.2. Participants

Seventeen elite women soccer players with a mean ± standard deviation age, height, body mass, and body mass index of 22.7 ± 6.3 years, 167.5 ± 5.6 cm, 60.7 ± 6.6 kg, and 21.6 ± 0.2 kg/m^2, respectively, participated in this study. Their experience as professional soccer players was 4.7 ± 2.2 years.

We estimated the power of the sample size using a post hoc *F*-test: the within-group factor in a repeated-measures MANOVA, according to statistical method analyzed. The analysis featured 94.2% of actual power, with a total of 17 women soccer players with a $p < 0.05$ and effect-size for 0.6, using G-Power [16].

The players belonged to a team that participated in the Portuguese BPI League during the 2019/20 in-season. The inclusion criteria were regular participation in most of the training sessions (80% of the weekly training sessions) and the completion of at least half the matches in the first half of the season [17], while the exclusion criteria were injury, illness, sickness, and/or non-performance of all the assessments. Due to the exclusion criteria, only sixteen women soccer players participated in the present study. The field positions of the players in the study consisted of one goalkeeper, three central defenders (CD), three wide defenders (WD), three central midfielders (CM), four wide midfielders (WM) and three strikers (ST).

Despite the different characteristics of the soccer field players, the goalkeeper was included in the analysis, since all the data collected for this player were similar to the squad average and the players' position values, and it was not detected as an outlier. All the participants were familiarized with the training protocols and the study design was carefully explained to the athletes. Written informed consent was obtained prior to the investigation.

A food frequency questionnaire to assess nutritional intake was applied over a 7 day period using a 24 h diet record, during the first week of the assessment 1 and during the last week of the assessment 3, in order for the players to verify their habits and food regimen routines.

The participants were instructed regarding portion sizes, supplements, food preparation aspects, and other aspects pertaining to an accurate recording of their energy intake. The records were reviewed for macronutrient composition and total energy intake [7]. All the participants were asked to maintain their normal diet throughout the study period.

The study was conducted according to the requirements of the Declaration of Helsinki and all the procedures were approved by the research Ethics Committee of the Polytechnic Institute of Santarém, Santarém, Portugal. All the subjects received their club's medical approval to participate in the study and were instructed not to take any medication during the study.

2.3. Procedures

The data were collected in weeks with only one match, which means that the team typically trained three days a week (match day minus (-); MD-5; MD-4; MD-2). This approach was used in a previous study [17]. During the period analyzed, a total of 57 training sessions and 13 matches occurred. The 57 training sessions were divided into 19 speed endurance sessions (e.g., long sprints, repeated sprints), 19 aerobic high-intensity sessions (e.g., interval training, medium-to-large sized games), and 19 ball-possession games and team/opponent tactics sessions. Figure 1 presents the timeline of the study.

In order to produce more specific information regarding training and match load, rated perceived exertion (RPE) and the duration of training sessions and matches were collected and presented in Table 1 to quantify training load. The data are presented by squad average between the different assessments. On match days (MD) only the average data for starters were included.

ASSESSMENTS

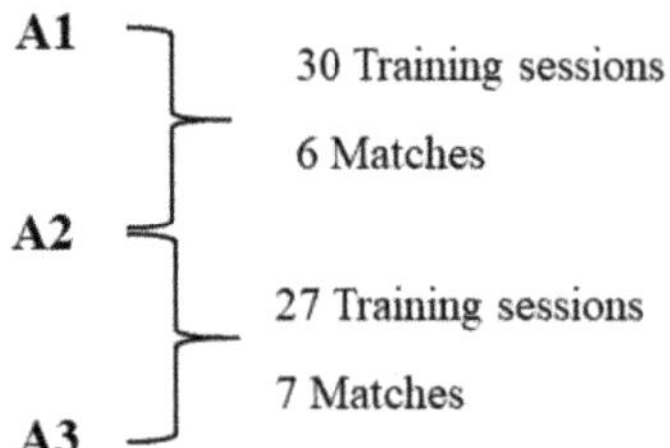

Figure 1. Timeline of the study. Legend: A1. Assessment 1; A2. Assessment 2; A3. Assessment 3.

Table 1. Training and match RPE and duration between the three assessments.

Periods	Variables	MD-5	MD-4	MD-3	MD-2	MD-1	MD
	RPE (au)	5.5	5.4	X	4.8	X	6.2
Between A1 and A2	Duration (min)	87	85	X	77	X	72
	s-RPE (au)	478.5	459	X	396.6	X	446.4
	RPE (au)	6.1	5.5	X	4.1	X	6.5
Between A2 and A3	Duration (min)	85	85	X	90	X	90
	s-RPE (au)	518.5	467.5	X	369	X	585

A1. Assessment 1; A2. Assessment 2; A3. Assessment 3; MD. Match-day; MD-5. Match minus five days to the match. respectively for -4, -3, -2, and -1. RPE. Rated perceived exertion; s-RPE. Session rated perceived exertion; au. Arbitrary units; min. Minutes. X indicates day off.

2.4. Anthropometric and Body Composition Assessment

Based on previous recommendations, the anthropometric and body composition measures were obtained with the subjects dressed in light clothing without shoes [18,19]. The participants were further asked to remove all objects that could interfere with the bioelectrical impedance assessment. The participants' weight and height were measured using a stadiometer with an incorporated scale (Seca 220, Hamburg, Germany) according to standardized procedures [20]. The body composition data were obtained with bioelectrical impedance analysis through Inbody S10 (model JMW140, Biospace Co, Ltd., Seoul, Korea), according to manufacturer's guidelines [21,22] and the recommendations of a previous study [23]. Eight electrodes were placed on eight tactile points (thumbs, middle fingers and ankles of both hands and feet, respectively) to perform the multi-segmental frequency analysis. Next, a total of 30 impedance measurements were obtained at frequencies 1, 5, 50, 250, 500, and 1000 kHz, respectively, from different segments of the body, such as the right and left arms, trunk, and right and left legs, respectively. Moreover, three different frequencies (5, 50, and 250 kHz) were used to collect the 15 reactance, PhA measurements from the right and left arms, trunk, and right and left legs, respectively. The variables collected were: body fat mass (BFM), soft lean mass (SLM), fat-free mass (FFM), intracellular water (ICW), extracellular water (ECW), total body water (TBW), phase angle (PhA, 50 Khz), ECW/TBW ratio, and ECW/ICW.

The measurements were carried out in the morning [18,24], in a room with an ambient temperature and relative humidity of 22–23 °C and 50–60%, respectively, after a minimum of 8 h of fasting and after the bladder was emptied. The participants adopted a supine position with their arms and legs abducted at a 45° angle, and the right hand and foot dorsal surfaces were cleaned with alcohol. After a 10 min rest in a room without noise, eight electrodes were placed on the cleaned surfaces and the measurements were performed. The subjects did not exercise or ingest caffeine or alcohol during the 12 h prior to the assessment and they were only assessed if they were in the luteal phase of ovulatory menstrual cycles. Otherwise, they waited for more days, until they were in the luteal phase.

All the assessments were performed by the same evaluator in order to minimize possible measurement errors [25].

2.5. Training and Match Load Quantification

Thirty minutes after the end of each training session and match, the players were asked to provide an RPE (0–10 scale) [26]. The players were prompted for their RPE individually using a custom-designed application on a portable computer tablet. They selected their RPE rating by touching the respective score on the tablet, which was then automatically saved under the player's profile. This method helped to minimize factors that may have influenced the player's RPE rating, such as peer pressure and replicating other players' ratings [27]. Next, the s-RPE was calculated, as in our previous studies, through the multiplication of the session duration by the RPE [28,29].

2.6. Statistical Procedures

Descriptive statistics (mean ± standard deviation) were performed for all the measurements. All the variables were checked for normality and homoscedasticity, respectively, using the Shapiro–Wilk and Levene tests. The MANOVA with repeated measures was performed for the variables that obtained normal distribution to compare the three assessments, with s-RPE being used as covariate. The value of $p \leq 0.05$ was established as significant and all the data were analyzed using SPSS version 22.0 (SPSS Inc., Chicago, IL, USA) for the Windows statistical software package. Furthermore, the change (%) was calculated between each comparison. The Cohen's d effect-size (ES) was performed to determine the effect magnitude through the difference of two means divided by the standard deviation from the data, and the following criteria were used: <0.2 = trivial, 0.2 to 0.6 = small effect, 0.6 to 1.2 = moderate effect, 1.2 to 2.0 = large effect, and >2.0 = very large [30].

3. Results

Table 2 summarizes the participants' characteristics by player position, while Table 3 showed comparisons between the three assessments for the squad average.

Table 2. Participant characteristics by player position in the three assessments.

Variables	Goalkeeper $n = 1$	Central Defender $n = 3$	Wide Defender $n = 3$	Central Midfielder $n = 3$	Wide Midfielder $n = 4$	Striker $n = 3$
			Assessment 1			
Body weight (kg)	64.0	71.0 ± 2.0	54.3 ± 3.8	59.3 ±9.2	53.5 ± 8.7	57 ± 1.0
Body fat mass (kg)	15.3	18.7 ± 2.3	12.4 ± 1.4	14.1 ± 5.4	11.1 ± 4.5	8.1 ± 2.0
Soft lean mass (kg)	45.9	49.1 ± 2.1	39.3 ± 2.3	42.5 ± 3.6	39.9 ± 5.1	46.0 ± 2.9
Fat free mass (kg)	48.7	52.3 ± 2.2	41.9 ± 2.4	45.2 ± 3.9	42.5 ± 5.4	48.9 ± 3.0
Intracellular Water (L)	22.4	23.8 ± 0.9	19.1 ± 1.3	20.6 ± 1.9	19.3 ± 2.4	22.4 ± 1.3
Extracellular Water (L)	13.2	14.4 ± 0.8	11.4 ± 0.6	12.5 ± 1.0	11.7 ± 1.5	13.3 ± 1.0
Total Body Water (L)	35.6	38.2 ± 1.7	35.5 ± 1.8	33.1 ± 2.8	31.0 ± 4.0	35.7 ± 2.3
Phase Angle (θ. 50 Khz)	6.8	6.0 ± 0.3	6.3 ± 0.6	6.3 ± 0.5	6.0 ±0.3	6.4 ± 0.3
			Assessment 2			
Body weight (kg)	67.0	69.3 ± 1.2	53.7 ± 3.2	58.0 ± 6.9	53.5 ± 7.9	57.0 ± 2.0
Body fat mass (kg)	15.8	14.1 ± 2.6	8.8 ± 4.2	10.2 ± 1.9	10.9 ± 3.7	6.7 ± 1.8
Soft lean mass (kg)	48.1	51.7 ± 2.1	42.0 ± 6.0	44.8 ± 4.9	39.9 ± 5.3	47.2 ± 3.5
Fat free mass (kg)	51.2	55.3 ± 2.3	44.9 ± 6.2	47.8 ± 5.1	42.6 ± 5.6	50.3 ± 3.7
Intracellular Water (L)	23.5	25.2 ± 1.1	20.5 ± 3.0	21.8 ± 2.4	19.4 ± 2.7	23.1 ± 1.7
Extracellular Water (L)	13.9	15.0 ± 0.6	12.1 ± 1.6	13.1 ± 1.3	11.6 ± 1.4	13.6 ± 1.1
Total Body Water (L)	37.4	40.1 ± 1.7	32.6 ± 4.6	34.8 ± 3.7	31.0 ± 4.1	36.7 ± 2.8
Phase Angle (θ. 50 Khz)	6.8	6.5 ± 0.6	7.8 ± 2.6	6.6 ± 0.7	6.2 ± 0.4	6.7 ± 0.2

Table 2. *Cont.*

Variables	Goalkeeper $n = 1$	Central Defender $n = 3$	Wide Defender $n = 3$	Central Midfielder $n = 3$	Wide Midfielder $n = 4$	Striker $n = 3$
			Assessment 3			
Body weight (kg)	67	69.0 ± 2.6	53 ± 4.4	57.0 ± 6.2	53.8 ± 7.4	59.0 ± 1.7
Body fat mass (kg)	15.4	12.1 ± 3.6	8.0 ± 2.8	12.2 ± 3.6	9.4 ± 3.1	8.7 ± 1.7
Soft lean mass (kg)	48.4	53.3 ± 5.4	42.2 ± 2.6	42.0 ± 2.6	41.5 ± 4.1	47.2 ± 2.9
Fat free mass (kg)	51.6	56.9 ± 5.6	45.0 ± 2.7	44.8 ± 2.9	44.4 ± 4.5	50.3 ± 2.9
Intracellular Water (L)	23.7	26.1 ± 2.7	20.6 ± 1.4	20.4 ± 1.3	20.1 ± 2.0	23.1 ± 1.4
Extracellular Water (L)	13.9	15.4 ± 1.4	12.1 ± 0.6	12.2 ± 0.7	2.3 ± 1.3	13.5 ± 0.8
Total Body Water (L)	37.6	41.4 ± 4.1	32.7 ± 1.9	32.6 ± 2.0	32.3 ± 3.3	33.6 ± 2.2
Phase Angle (θ. 50 Khz)	7.4	7.1 ± 0.6	7.4 ± 0.5	6.9 ± 0.3	6.6 ± 0.2	6.9 ± 0.3

Table 3. Comparisons between assessments by squad average ($n = 17$).

Variables	A1	A2	A3	Change % (A1–A2)	Change % (A2–A3)	Change % (A1–A3)
Body weight (kg)	58.74 ± 2.15	58.30 ± 1.97	58.30 ± 1.94	−0.8	0.0	−0.8
Body fat mass (kg)	13.11 ± 1.87 a	10.77 ± 0.94	10.38 ± 0.87	−21.7	−3.8	−26.3
Soft lean mass (kg)	42.87 ± 1.20 a	44.52 ± 1.44	44.91 ± 1.42	3.7	0.9	4.5
Fat free mass (kg)	45.63 ± 1.27 a.c	47.52 ± 1.53	47.92 ± 1.51	4.0	0.8	4.8
Intracellular Water (L)	20.79 ± 0.58 a.c	21.71 ± 0.72	21.88 ± 0.71	4.2	0.8	5.0
Extracellular Water (L)	12.53 ± 0.35	12.88 ± 0.41	13.00 ± 0.39	2.7	0.9	3.6
Total Body Water (L)	33.32 ± 0.93 a	34.59 ± 1.12	34.88 ± 1.09	3.7	0.8	4.5
ECW/TBW	0.38 ± 0.001 a.c	0.37 ± 0.001 b	0.37 ± 0.001	−2.7	0.0	−2.7
ECW/ICW	0.60 ± 0.003 a	0.59 ± 0.003	0.59 ± 0.004	−1.7	0.0	−1.7
Phase Angle (θ. 50 Khz)	6.26 ± 0.11 c	6.67 ± 0.31	6.99 ± 0.10	6.1	4.6	10.4

A1. Assessment 1; A2. Assessment 2; A3. Assessment 3; ECW. Extracellular water; ICW. Intracellular water; TBW. Total body water. The symbol a denotes significant difference between A1 and A2 ($p < 0.05$). The symbol b denotes significant difference between A2 and A3 ($p < 0.05$). The symbol c denotes significant difference between A1 and A3 ($p < 0.05$).

After performing ANCOVA with the session's rated perceived exertion (s-RPE) as the covariate, no linear interaction was demonstrated between this variable and any of the other body composition variables ($p > 0.05$). Table 2 shows significant differences between A1 and A2 with moderate to very large effect, namely, BFM ($p = 0.029$; ES = 1.58), SLM ($p = 0.018$; ES = 1.24), FFM ($p = 0.010$; ES = 1.34), ICW ($p = 0.007$; ES = 1.41), TBW ($p = 0.018$; ES = 1.04), ECW/TBW ($p = 0.002$; ES = 10.00), and ECW/ICW ($p = 0.022$; ES = 3.33).

In addition, there was only a significant difference with very large effect between A2 and A3, for ECW/TBW ($p = 0.001$; ES = 3.33).

Finally, there were significant differences with large to very large effect between A1 and A3 for BFM ($p = 0.029$; ES = 1.87), FFM ($p = 0.045$; ES = 1.51), ICW ($p = 0.049$; ES = 1.68), ECW/TBW ($p = 0.013$; ES = 10.00), and PhA ($p = 0.001$; ES= 6.64).

4. Discussion

In this study, we aimed to identify changes in the body composition of elite women soccer players during in-season through BIA. Our main findings showed improvements in body composition, namely decreased BFM, increased FFM, and increased PhA; and a better fluid distribution was observed, especially from the first to the last assessment. However, no significant differences were noted between A2 and A3, except for ECW/TBW.

On one hand, BFM has been shown to exert a negative influence in athletes' performance [5]. On the other hand, FFM has been associated with performance improvements [5]. In our study, the athletes showed a significant decrease in BFM and an increase in FFM. These results are similar to those reported in another study [31], which assessed athletes' body composition in 5 time-points during the in-season. Regarding BFM, athletes presented mean values similar to those found by other authors [32] that assessed body composition changes pre-to-post-season in women soccer players. However, the authors found that the soccer players lost lean mass tissue over the competitive season that was not recovered during the off-season [32]. These results may be attributed to overtraining or negative energy balance.

Concerning PhA, the present female soccer players showed a mean value of $6.26 \pm 0.11°$ in A1, $6.67 \pm 0.31°$ in A2, and $6.99 \pm 0.10°$ in A3. All these values are similar to those found in other studies conducted on women athletes or active young populations [33–35]. Furthermore, the values obtained in this study were slightly lower in A1 and A2 compared to those obtained in a study of healthy adult non-athletes [36]. Moreover, PhA has been related to cellular health and integrity [37]. For example, muscle injuries can cause a reduction in PhA which can provoke cell membrane disruption [38,39], which has also been related to body composition [33,40,41]. For instance, FFM is directly related with PhA [41]. Indeed, as FFM increased in these athletes, it seems plausible that PhA also increased. An improvement in PhA can be an indicator of good health and cellular integrity and functionality regarding the level of hydration [34]. Another application of PhA is related to cellular energy levels, so the low phase angle is consistent with an inability of cells to store energy, as well as being an indication of breakdown in the selective permeability of cellular membranes. A high PhA is consistent with large quantities of intact cell membranes and body cell mass [42].

Regarding TBW and its compartments, the importance of TBW and ICW in increasing performance in athletes is clear [7–9]. The increment in ICW and TBW in the present study is in line with a previous study that used resistance training in healthy and young adults [35]. In this regard, soccer is characterized by high intensity bouts of activities and movements. Glycogen is an essential substrate during high intensity sports [43]. Therefore, some explanations could be related to cellular hydration by increasing the glycogen storage, since glycogen features great osmotic power (each gram of glycogen is stored in human muscle with at least 3 g of water) [43]. These results are very important for athletes, since ICW content may stimulate pathways that increase protein synthesis [44,45]. ECW did not show any change during the in-season. Furthermore, the ECW/ICW ratio has been used as an indicator of fluid distribution in athletes [7–9,33]. Two recent studies [33,34] found values of 0.7 ± 0.1 in women athletes. In our study, the soccer athletes demonstrated mean values of 0.60 in A1, 0.59 in A2, and 0.59 in A3. Lower values of ECW/ICW have been found in athletes has and they have been associated with improved performance [7].

As mentioned earlier, when A2 andA3 were compared, no significant differences were found. These findings could be attributed to the increased training load in the beginning of the in-season that is generally found in soccer teams [46]. The higher training load resulted in body composition improvements in this early phase (between A1 and A2) that were followed by an adaptation in the second phase of the study (between A2 and A3), causing a maintenance of the body composition variables (considering that nutritional intake was controlled). This is important to highlight because in fact training load was higher between A2 and A3 without, however, changing any body composition variable.

A relevant finding that should be highlighted regards s-RPE. Through the analysis conducted in the present study, no interaction was observed between s-RPE and any body composition variables, which means that RPE can be dissociated from the physiological process through different psychological mechanisms [47]. As mentioned in previous studies, it seems that RPE was a simplification of the perceived psychophysiological exertion. Consequently, the use of this measure alone did not conclusively capture different sensations and experience of training sessions [47,48]. Furthermore, RPE was collected 30 min after the training sessions and that value included the entire session. This means that there could be some possible variation during training sessions in different exercises, as suggested by Ferraz et al. [48], that were not controlled in this study. This explanation may help to explain the non-interaction found regarding this variable in this study. It also reinforces the use of additional variables in training load monitoring, such as distances covered at different intensities, accelerations, decelerations, player load and metabolic power.

Scientific research on women soccer athletes is scarce [10,11,49], especially at the elite level, and to the best of our knowledge, this is one of the first studies to include several variables in order to assess anthropometric and body composition in elite women soccer players from a Portuguese BPI Ligue team. However, the sample size was derived from one team only and, therefore, future studies are required to generalize the findings.

Another interesting finding was related to the goalkeeper analysis, which showed that s-RPE and different body composition variables were similar regardless of player position and the squad average values. However, future studies are required to confirm this finding, since only one goalkeeper is an insufficient sample size from which to draw definite conclusions. Furthermore, more players for each position are required for an analysis across player positions.

Despite the importance of these results, and despite the use of tetra-polar and multi-frequency bioimpedance equipment, such as InBody S10, to assess body composition and fluid distribution, we should address, as a limitation of this study, the use of a non-considered reference method. Another limitation was the fact that it was not possible to make comparisons among athletes of different field positions, as this would reduce the sampling power. Finally, and despite the fact that no differences were found in nutritional intake, this assessment was performed through a questionnaire at two time points, which should be better addressed in future studies. Even so, this study represents the actual training routine followed by the specific team analyzed. Therefore, more research is needed with larger numbers regarding soccer players and teams over an all-season period.

5. Conclusions

Coaches, physicians, nutritionists, and exercise physiologists should ensure they provide gender-specifications for optimizing performance. This study highlights information on the essential characteristics of successful women's' soccer team performance at three time-points throughout the sport season. For instance, the study showed that although some players may have performed different field roles and positions, their body composition characteristics improved over the season, which reveals that nutritional habits were controlled and, consequently, the intensity of training and matches did not affect the body composition variables.

This study presents a report using body composition data and internal training load simultaneously, which can be used as a reference for better body composition, training load and performance management for coaches and their staff. However, we recommend that future studies include a full season and other training load measures, such as global positioning systems, to amplify the present findings.

Author Contributions: Conceptualization, R.O.; methodology, R.O., J.P.B. and F.M.C.; software, R.O., A.M. and R.F.(Renato Fernandes); validation, R.O. and J.P.B.; formal analysis, R.O. and J.P.B.; investigation, R.O., R.F. (Ruben Francisco), A.M., J.P.B., H.N., F.M.C. and R.F. (Renato Fernandes); resources, R.O., R.F. (Renato Fernandes), A.M., J.P.B., H.N., F.M.C. and R.F. (Renato Fernandes); data curation, A.M. and R.F. (Renato Fernandes); writing—original draft preparation, R.O., R.F. (Ruben Francisco), A.M., J.P.B., H.N., F.M.C. and R.F. (Renato Fernandes); writing—review and editing, R.O., R.F. (Ruben Francisco), A.M., J.P.B. and F.M.C.; visualization, R.O., R.F. (Renato Fernandes), J.P.B. and F.M.C.; supervision, R.O. and F.M.C.; project administration, R.O., J.P.B. and R.F. (Renato Fernandes); funding acquisition, R.O. and J.P.B. All authors have read and agreed to the published version of the manuscript.

Funding: This research was funded by the Portuguese Foundation for Science and Technology, I.P., Grant/Award Number UIDP/04748/2020. The funders had no role in the design of the study; in the collection, analyses, or interpretation of data; in the writing of the manuscript, or in the decision to publish the results.

Institutional Review Board Statement: The study was conducted according to the guidelines of the Declaration of Helsinki and approved by the Ethics Committee of Polytechnic Institute of Santarém (252020Desporto).

Informed Consent Statement: Written informed consent was obtained from the participants to publish this paper.

Data Availability Statement: The data presented in this study are available on request from the corresponding author.

Acknowledgments: The authors would like to thank the team's coaches and players for their cooperation during all data collection procedures.

Conflicts of Interest: The authors declare no conflict of interest.

References

1. Turner, A.N.; Stewart, P.F. Strength and conditioning for soccer players. *Strength Cond. J.* **2014**, *36*, 1–13. [CrossRef]
2. Sutton, L.; Scott, M.; Wallace, J.; Reilly, T. Body composition of English Premier League soccer players: Influence of playing position, international status, and ethnicity. *J. Sports Sci.* **2009**, *27*, 1019–1026. [CrossRef]
3. Silva, A.M. Structural and functional body components in athletic health and performance phenotypes. *Eur. J. Clin. Nutr.* **2019**, *73*, 215–224. [CrossRef] [PubMed]
4. Rienzi, E.; Drust, B.; Reilly, B.; Carter, J.E.L.; Martin, A. Investigation of anthropometric and work-rate profiles of elite South American international soccer players. *J. Sports Med. Phys. Fit.* **2000**, *40*, 162.
5. Silvestre, R.; West, C.; Maresh, C.M.; Kraemer, W.J. Body Composition And Physical Performance In Men's Soccer: Astudy Of A National Collegiate Athletic Association Division Iteam. *J. Strength Cond. Res.* **2006**, *20*, 177–183. [CrossRef]
6. Collins, J.; Maughan, R.J.; Gleeson, M.; Bilsborough, J.; Jeukendrup, A.; Morton, J.P.; Phillips, S.M.; Armstrong, L.; Burke, L.M.; Close, G.L.; et al. UEFA expert group statement on nutrition in elite football. Current evidence to inform practical recommendations and guide future research. *Br. J. Sports Med.* **2021**, *55*, 416. [CrossRef]
7. Silva, A.M.; Matias, C.N.; Santos, D.A.; Rocha, P.M.; Minderico, C.S.; Sardinha, L.B. Increases in intracellular water explain strength and power improvements over a season. *Int. J. Sports Med.* **2014**, *35*, 1101–1105. [CrossRef]
8. Silva, A.M.; Fields, D.A.; Heymsfield, S.B.; Sardinha, L.B. Body composition and power changes in elite judo athletes. *Int. J. Sports Med.* **2010**, *31*, 737–741. [CrossRef] [PubMed]
9. Silva, A.M.; Fields, D.A.; Heymsfield, S.B.; Sardinha, L.B. Relationship between changes in total-body water and fluid distribution with maximal forearm strength in elite judo athletes. *J. Strength Cond. Res.* **2011**, *25*, 2488–2495. [CrossRef]
10. Matković, B.R.; Mišigoj-Duraković, M.; Matković, B.; Janković, S.; Ružić, L.; Leko, G.; Miran, K. Morphological differences of elite Croatian soccer players according to the team position. *Coll. Antropol.* **2003**, *27*, 167–174. [PubMed]
11. Cárdenas-Fernández, V.; Chinchilla-Minguet, J.L.; Castillo-Rodríguez, A. Somatotype and body composition in young soccer players according to the playing position and sport success. *J. Strength Cond. Res.* **2019**, *33*, 1904–1911. [CrossRef]
12. da Silva, C.D.; Bloomfield, J.; Marins, J.C.B. A review of stature, body mass and maximal oxygen uptake profiles of U17, U20 and first division players in Brazilian soccer. *J. Sports Sci. Med.* **2008**, *7*, 309. [PubMed]
13. Owen, A.L.; Lago-Peñas, C.; Dunlop, G.; Mehdi, R.; Chtara, M.; Dellal, A. Seasonal body composition variation amongst elite european professional soccer players: An approach of talent identification. *J. Hum. Kinet.* **2018**, *62*, 177–184. [CrossRef]
14. Roelofs, E.; Bockin, A.; Bosch, T.; Oliver, J.; Bach, C.W.; Carbuhn, A.; Stanforth, P.R.; Dengel, D.R. Body Composition of National Collegiate Athletic Association (NCAA) Division I Female Soccer Athletes through Competitive Seasons. *Int. J. Sports Med.* **2020**, *41*, 766–770. [CrossRef] [PubMed]
15. Fanchini, M.; Ferraresi, I.; Petruolo, A.; Azzalin, A.; Ghielmetti, R.; Schena, F.; Impellizzeri, F.M. Is a retrospective RPE appropriate in soccer? Response shift and recall bias. *Sci. Med. Footb.* **2017**, *1*, 53–59. [CrossRef]
16. Faul, F.; Erdfelder, E.; Lang, A.-G.; Buchner, A. G* Power 3: A flexible statistical power analysis program for the social, behavioral, and biomedical sciences. *Behav. Res. Methods* **2007**, *39*, 175–191. [CrossRef] [PubMed]
17. Oliveira, R.; Brito, J.P.; Loureiro, N.; Padinha, V.; Nobari, H.; Mendes, B. Will Next Match Location Influence External and Internal Training Load of a Top-Class Elite Professional European Soccer Team? *Int. J. Environ. Res. Public Health* **2021**, *18*, 5229. [CrossRef]
18. Arazi, H.; Mirzaei, B.; Nobari, H. Anthropometric profile, body composition and somatotyping of national Iranian cross-country runners. *Turk. J. Sport Exerc.* **2015**, *17*, 35–41. [CrossRef]
19. Nobari, H.; Polito, L.F.T.; Clemente, F.M.; Pérez-Gómez, J.; Ahmadi, M.; Garcia-Gordillo, M.Á.; Garcia-Gordillo, M.Á.; Silva, A.F.; Adsuar, J.C. Relationships between training workload parameters with variations in anaerobic power and change of direction status in elite youth soccer players. *Int. J. Environ. Res. Public Health* **2020**, *17*, 7934. [CrossRef] [PubMed]
20. Lohman, T.G.; Roche, A.F.; Martorell, R. *Anthropometric Standardization Reference Manual*, 1st ed.; Human Kinetics Books: Champaign, IL, USA, 1988.
21. Yang, E.M.; Park, E.; Ahn, Y.H.; Choi, H.J.; Kang, H.G.; Cheong, H.I.; Ha, I.S. Measurement of fluid status using bioimpedance methods in Korean pediatric patients on hemodialysis. *J. Korean Med. Sci.* **2017**, *32*, 1828. [CrossRef]
22. Buckinx, F.; Reginster, J.-Y.; Dardenne, N.; Croisiser, J.-L.; Kaux, J.-F.; Beaudart, C.; Slomian, J.; Bruyère, O. Concordance between muscle mass assessed by bioelectrical impedance analysis and by dual energy X-ray absorptiometry: A cross-sectional study. *BMC Musculoskelet. Disord.* **2015**, *16*, 60. [CrossRef] [PubMed]

23. Martins, A.D.; Oliveira, R.; Brito, J.P.; Costa, T.; Ramalho, F.; Pimenta, N.; Santos-Rocha, R. Phase angle cutoff value as a marker of the health status and functional capacity in breast cancer survivors. *Physiol. Behav.* **2021**, *235*, 113400. [CrossRef]
24. Rahmat, A.J.; Arsalan, D.; Bahman, M.; Hadi, N. Anthropometrical profile and bio-motor abilities of young elite wrestlers. *Phys. Educ. Stud.* **2016**, *20*, 63–69. [CrossRef]
25. Steward, A.; Marfell-Jones, M.; Olds, T.; de Ridder, H. *International Standards for Anthropometric Assessment*; International Society for the Advancement of Kinanthropometry: Lower Hutt, New Zealand, 2011.
26. Borg, G. Perceived exertion as an indicator of somatic stress, Scand. *J. Rehabil. Med.* **1970**, *2*, 92–98.
27. Burgess, D.; Drust, B.; Williams, M. Developing a physiology-based sports science support strategy in the professional game. *Sci. Soccer Dev. Elite Perform.* **2013**, *13*, 372–389.
28. Foster, C. Monitoring training in athletes with reference to overtraining syndrome. *Occup. Health Ind. Med.* **1998**, *4*, 189. [CrossRef]
29. Foster, C.; Florhaug, J.A.; Franklin, J.; Gottschall, L.; Hrovatin, L.A.; Parker, S.; Pamela, D.; Christopher, D. A new approach to monitoring exercise training. *J. Strength Cond. Res.* **2001**, *15*, 109–115.
30. Hopkins, W.G. Spreadsheets for analysis of controlled trials, with adjustment for a subject characteristic. *Sportscience* **2006**, *10*, 46–50.
31. Peart, A.N.; Nicks, C.R.; Mangum, M.; Tyo, B.M. Evaluation of seasonal changes in fitness, anthropometrics, and body composition in collegiate division II female soccer players. *J. Strength Cond. Res.* **2018**, *32*, 2010–2017. [CrossRef]
32. Minett, M.M.; Binkley, T.B.; Weidauer, L.A.; Specker, B.L. Changes in body composition and bone of female collegiate soccer players through the competitive season and off-season. *J. Musculoskelet. Neuronal Interact.* **2017**, *17*, 386.
33. Francisco, R.; Matias, C.N.; Santos, D.A.; Campa, F.; Minderico, C.S.; Rocha, P.; Heymsfield, S.B.; Lukaski, H.; Sardinha, L.B.; Silva, A.M. The Predictive Role of Raw Bioelectrical Impedance Parameters in Water Compartments and Fluid Distribution Assessed by Dilution Techniques in Athletes. *Int. J. Environ. Res. Public Health* **2020**, *17*, 759. [CrossRef] [PubMed]
34. Marini, E.; Campa, F.; Buffa, R.; Stagi, S.; Matias, C.N.; Toselli, S.; Sardinha, L.B.; Silva, A.M. Phase angle and bioelectrical impedance vector analysis in the evaluation of body composition in athletes. *Clin. Nutr.* **2019**, *39*, 447–454. [CrossRef]
35. Ribeiro, A.S.; Avelar, A.; Santos, L.D.; Silva, A.M.; Gobbo, L.A.; Schoenfeld, B.J.; Sardinha, E.S.C. Hypertrophy-type Resistance Training Improves Phase Angle in Young Adult Men and Women. *Int. J. Sports Med.* **2017**, *38*, 35–40. [CrossRef]
36. Barbosa-Silva, M.C.; Barros, A.J.; Wang, J.; Heymsfield, S.B.; Pierson, R.N., Jr. Bioelectrical impedance analysis: Population reference values for phase angle by age and sex. *Am. J. Clin. Nutr.* **2005**, *82*, 49–52. [CrossRef]
37. Norman, K.; Stobaus, N.; Zocher, D.; Bosy-Westphal, A.; Szramek, A.; Scheufele, R.; Smoliner, C.; Pirlich, M. Cutoff percentiles of bioelectrical phase angle predict functionality, quality of life, and mortality in patients with cancer. *Am. J. Clin. Nutr.* **2010**, *92*, 612–619. [CrossRef] [PubMed]
38. Nescolarde, L.; Yanguas, J.; Lukaski, H.; Alomar, X.; Rosell-Ferrer, J.; Rodas, G. Localized bioimpedance to assess muscle injury. *Physiol. Meas.* **2013**, *34*, 237–245. [CrossRef]
39. Nescolarde, L.; Yanguas, J.; Medina, D.; Rodas, G.; Rosell-Ferrer, J. Assessment and follow-up of muscle injuries in athletes by bioimpedance: Preliminary results. *Conf. Proc. IEEE Eng. Med. Biol. Soc.* **2011**, *2011*, 1137–1140.
40. Norman, K.; Wirth, R.; Neubauer, M.; Eckardt, R.; Stobaus, N. The bioimpedance phase angle predicts low muscle strength, impaired quality of life, and increased mortality in old patients with cancer. *J. Am. Med. Dir. Assoc.* **2015**, *16*, 173-e17. [CrossRef] [PubMed]
41. Gonzalez, M.C.; Barbosa-Silva, T.G.; Bielemann, R.M.; Gallagher, D.; Heymsfield, S.B. Phase angle and its determinants in healthy subjects: Influence of body composition. *Am. J. Clin. Nutr.* **2016**, *103*, 712–716. [CrossRef] [PubMed]
42. Mascherini, G.; Gatterer, H.; Lukaski, H.; Burtscher, M.; Galanti, G. Changes in hydration, body-cell mass and endurance performance of professional soccer players through a competitive season. *J. Sports. Med. Phys. Fit.* **2015**, *55*, 749–755.
43. Kenney, W.L.; Wilmore, J.H.; Costill, D.L. *Physiology of Sport and Exercise*, 6th ed.; Human Kinetics Books: Champaign, IL, USA, 2015.
44. Haussinger, D.; Lang, F.; Gerok, W. Regulation of cell function by the cellular hydration state. *Am. J. Physiol. Endocrinol. Metab.* **1994**, *267*, E343–E355. [CrossRef]
45. Haussinger, D.; Roth, E.; Lang, F.; Gerok, W. Cellular hydration state: An important determinant of protein catabolism in health and disease. *Lancet* **1993**, *341*, 1330–1332. [CrossRef]
46. Mara, J.K.; Thompson, K.G.; Pumpa, K.L.; Ball, N.B. Periodization and physical performance in elite female soccer players. *Int. J. Sports Physiol. Perform.* **2015**, *10*, 664–669. [CrossRef]
47. Renfree, A.; Martin, L.; Micklewright, D.; Gibson, A.S.C. Application of decision-making theory to the regulation of muscular work rate during self-paced competitive endurance activity. *Sports Med.* **2014**, *44*, 147–158. [CrossRef]
48. Ferraz, R.; Goncalves, B.; Coutinho, D.; Marinho, D.A.; Sampaio, J.; Marques, M.C. Pacing behaviour of players in team sports: Influence of match status manipulation and task duration knowledge. *PLoS ONE* **2018**, *13*, e0192399. [CrossRef]
49. Julian, R.; Skorski, S.; Hecksteden, A.; Pfeifer, C.; Bradley, P.S.; Schulze, E.; Meyer, T. Menstrual cycle phase and elite female soccer match-play: Influence on various physical performance outputs. *Sci. Med. Footb.* **2020**, *5*, 97–104. [CrossRef]

International Journal of
Environmental Research and Public Health

MDPI

Article

In-Season Internal Load and Wellness Variations in Professional Women Soccer Players: Comparisons between Playing Positions and Status

Renato Fernandes [1,2,3,*], João Paulo Brito [1,2,4], Luiz H. Palucci Vieira [5], Alexandre Duarte Martins [1,2,6], Filipe Manuel Clemente [7,8], Hadi Nobari [9,10,11,12], Victor Machado Reis [3,4] and Rafael Oliveira [1,2,4,*]

[1] Sports Science School of Rio Maior–Polytechnic Institute of Santarém, 2040-413 Rio Maior, Portugal; jbrito@esdrm.ipsantarem.pt (J.P.B.); alexandremartins@esdrm.ipsantarem.pt (A.D.M.)
[2] Life Quality Research Centre, 2040-413 Rio Maior, Portugal
[3] University of Trás-os-Montes e Alto Douro, 5001-801 Vila Real, Portugal; victormachadoreis@gmail.com
[4] Research Centre in Sport Sciences, Health Sciences and Human Development, 5001-801 Vila Real, Portugal
[5] Graduate Program in Movement Sciences, MOVI-LAB Human Movement Research Laboratory, Physical Education Department, School of Sciences, UNESP São Paulo State University, Bauru 17033-360, Brazil; luiz.palucci@unesp.br
[6] Comprehensive Health Research Centre (CHRC), Departamento de Desporto e Saúde, Escola de Saúde e Desenvolvimento Humano, Universidade de Évora, Largo dos Colegiais, 7004-516 Évora, Portugal
[7] Escola Superior Desporto e Lazer, Instituto Politécnico de Viana do Castelo, Rua Escola Industrial e Comercial de Nun'Álvares, 4900-347 Viana do Castelo, Portugal; filipe.clemente5@gmail.com
[8] Instituto de Telecomunicações, Delegação da Covilhã, 1049-001 Lisboa, Portugal
[9] Department of Physical Education and Sports, University of Granada, 18010 Granada, Spain; hadi.nobari1@gmail.com
[10] HEME Research Group, Faculty of Sport Sciences, University of Extremadura, 10003 Cáceres, Spain
[11] Department of Exercise Physiology, Faculty of Educational Sciences and Psychology, University of Mohaghegh Ardabili, Ardabil 56199-11367, Iran
[12] Sports Scientist, Sepahan Football Club, Isfahan 81887-78473, Iran
* Correspondence: rfernandes@esdrm.ipsantarem.pt (R.F.); rafaeloliveira@esdrm.ipsantarem.pt (R.O.)

Citation: Fernandes, R.; Brito, J.P.; Vieira, L.H.P.; Martins, A.D.; Clemente, F.M.; Nobari, H.; Reis, V.M.; Oliveira, R. In-Season Internal Load and Wellness Variations in Professional Women Soccer Players: Comparisons between Playing Positions and Status. *Int. J. Environ. Res. Public Health* **2021**, *18*, 12817. https://doi.org/10.3390/ijerph182312817

Academic Editor: Paul B. Tchounwou

Received: 30 October 2021
Accepted: 2 December 2021
Published: 5 December 2021

Publisher's Note: MDPI stays neutral with regard to jurisdictional claims in published maps and institutional affiliations.

Abstract: The internal intensity monitoring in soccer has been used more in recent years in men's football; however, in women's soccer, the existing literature is still scarce. The aims of this study were threefold: (a) to describe the weekly variations of training monotony, training strain and acute: chronic workload ratio through session Rated Perceived Exertion (s-RPE); (b) to describe weekly variations of Hooper Index [stress, fatigue, Delayed Onset Muscle Soreness (DOMS) and sleep]; and (c) to compare those variations between playing positions and player status. Nineteen players (24.1 ± 2.7 years) from a Portuguese BPI League professional team participated in this study. All variables were collected in a 10-week in-season period with three training sessions and one match per week during the 2019/20 season. Considering the overall team, the results showed that there were some associations between Hooper Index categories and s-RPE like stress or fatigue (0.693, $p < 0.01$), stress or DOMS (0.593, $p < 0.01$), stress or s-RPE (−0.516, $p < 0.05$) and fatigue or DOMS (0.688, $p < 0.01$). There were no differences between all parameters in playing positions or player status. In conclusion, the study revealed that higher levels of fatigue and DOMS occur concurrently with better nights of sleep. Moreover, any in-season variations concerning internal load and perceived wellness seems independent of position or status in outfield players. The data also showed that the higher the players' reported stress, the lower the observed s-RPE, thus possibly indicating a mutual interference of experienced stress levels on the assimilation of training intensity by elite women soccer players.

Keywords: muscle soreness; female; stress; fatigue; sleep; perceived exertion; training monotony; training strain

1. Introduction

Load/intensity monitoring is a well-implemented practice in team sports that guides the coach's interventions through a better understanding of the impact of training stimulus on players [1,2]. Monitoring consists of using a given instrument or technique that allows us to track the intensity of exercise in each player. Thus, a wide vision of how the player is coping with the training process is implementing a strategy in which training intensity monitoring is complemented by wellness and readiness monitoring [3]. This integrated approach, also known as athlete's monitoring allows understanding of the mechanisms related to training stimulus and recovery, thus providing some information about how the training periodization is actually done while helping coaches to quickly identify the individual responses of players to stressful situations while monitoring their wellness [4].

Due to the evident variations of a competitive schedule and as a part of the training methodology, it is expected to have in seasonal variations of training intensity (relative or absolute) in players [5]. Training demands are closely related to the training goals and structure of exercise; namely organization, quality, and quantity [2]. Thus, training intensity can be understood as an input variable constrained by the organization of training activities to elicit a given training response in accordance with the coach's expectations [2]. Training intensity monitoring can be organized in two main dimensions: external and internal. The contextualization of these two dimensions is important, since internal responses (i.e., psychophysiological responses) are closely related with the physical demands imposed by a given training drill.

The external demands represent the immediate player's physical responses to the organization, quality, and quantity of exercise (training plan) [2]. In the case of team sports, such physical responses are commonly analyzed using microelectromechanical systems or optical systems that provide estimated values related to distance-based, accelerometry-based, and combined variables [1]. These devices have been used not only in soccer, but in different women's team sports (e.g., rugby, volleyball, handball) in which distance-based and accelerometry-based measures are important to understand the dynamics of training in an heterogenous groups [6,7]. Currently, the use of microelectromechanical systems (e.g., global positioning systems, local positioning systems or inertial measurement units) allows us to individualize the understanding of intensity demands in women sports such as rugby [8] or soccer [9]. While the external intensity is the acute physical response to the session training plan, the internal demands can be understood as the psychophysiological response to the external demands [10]. This means that, although there are possible similarities in external load, the internal load can be considerably different between two players. Internal responses can be constrained by the individual characteristics of the players, training status, psychological status, health, nutrition, environment, or genetics [2]. Thus, it is reasonable to predict that the same training plan, with possible similar external loads, can provide different internal loads in players. While internal demands represent the acute psychophysiological responses to the exercise, it is expectable that the consistency of the internal training demands across time leads to adaptations in the physiological levels of players with natural variations in the training outcomes [11–13].

Internal intensity is a crucial part of the training monitoring processes. Among different possibilities for monitoring the internal measures (e.g., heart rate monitors, respiratory gas analyzer, and blood lactate), the Rated Perceived Exertion (RPE) is the easiest instrument to apply, since it ensures a valid, reliable, and sensitive approach to quantify and qualify the internal load while using a simple questionnaire [14,15]. Moreover, the final score obtained by the RPE questionnaire can provide useful information to estimate the internal intensity (namely, multiplying the RPE score by the time of training in minutes) [16], or even using this RPE-based training load to estimate the organization of the training, calculating the variability of the load applied in the week (e.g., using an equation to estimate the Training Monotony [TM]) [17], the progression of load across the weeks (e.g., using the Acute: Chronic Workload Ratio [ACWR] to identify a measure of increase of decrease of a load of a week in comparison to the previous one) [18] or identify the Training Strain [TS] of loads during a week (e.g., multiplying the training monotony by the acute load of the week) [17].

Although some research claims cause–effect consequences of specific intensity measures and injury occurrence [19], this has been dismissed since the absence of quality data and proper methodological approaches to prove that [20–22]. However, these intensity measures are still valuable for guiding coaches to understand the dynamics of stimulus and the impact of the training plans on the actual responses of the players as recently shown in women soccer players [23]. The main evidence about internal intensity variations across the season, while using RPE-based measures, suggests that pre-season is the period in which the internal intensity accumulated in the week and the intensity measures of monotony and strain are typically greater than in the in-season periods [24]. This can be related with the higher external demands occurring in such a period [25], as a consequence of the typical strategy of increasing the volume and frequency of training sessions to provide a higher stress on the player's organism before starting competition.

Although no strong relationships between training intensity and wellness are identified [26], wellness can be related to different variables, thus we can assist with variations across the season. In the context of soccer monitoring, wellness is quantified and qualified by the Delayed Onset Muscle Soreness (DOMS), sleep quality, fatigue, stress and mood, partially justifying these outcomes by the proposal of Hooper [27] which were updated by MacLean in 2010 [28]. Wellness is currently analyzed using questionnaires and those seem to be sensitive to variations within and between weeks (or periods). As an example, comparisons between pre-season and in-season revealed that pre-season was more strenuous and exhausting for players than the in-season period regarding the wellness variables that were inspected [29].

The main evidence on training monitoring and wellness in soccer has been related to men. However, women soccer players and training monitoring is still growing and due to the natural biological differences, more research to understand the mechanisms of how they cope with training process is needed. Thus, descriptive studies, namely cohorts, are still valuable for characterizing the reality of the training process in women's soccer. Based on that reason, the purposes of this study were threefold: (a) to describe the weekly variations of TM, TS and ACWR through s-RPE; (b) to describe weekly variations of stress, fatigue, DOMS and sleep; and (c) to compare those variations between playing positions and player status.

2. Materials and Methods

2.1. Subjects

Considering several studies conducted with small sample sizes [5,11–13,23,25,29–36], 19 professional female soccer players participated in this study (24.1 ± 2.7 years, 164.3 ± 4.2 cm, 58.5 ± 8.2 kg). The players belong to a team that participated in the Portuguese BPI League in the 2019/20 season. Based on player status, they were divided into two groups: starters ($n = 11$) and non-starters ($n = 8$). Additionally, the playing positions were divided into five defenders, five central midfielders, four wide midfielders and five strikers. The inclusion criteria included regular participation in most of the training sessions (80% of weekly training sessions), while the exclusion criteria included lack of player information, illness and/or injury for two consecutive weeks. Goalkeepers were excluded from the study. The criteria to define starters and non-starters were assessed week by week against a player's attendance time at the match and training sessions, and to be considered a starter, a player had to complete at least 60 min in three consecutive matches; players who did not achieve this duration were considered non-starters [34]. All participants were familiarized with the training protocols and signed informed consent prior to the investigation. This study was conducted according to the requirements of the Declaration of Helsinki and was approved by the Ethics Committee of Polytechnic Institute of Santarém (252020 Desporto).

2.2. Design

Training and match data were collected over a 10-week in-season period (between October and December) with three training sessions and one match per week. For the purposes of the present study, all the sessions carried out as the main team sessions were considered. This refers to training sessions in which both the starting and non-starting players trained together. Data from rehabilitation or additional training sessions of recuperation were excluded. This means that sessions after the match day were included whenever both starters and non-starters trained together, but other kinds of recovery training were excluded. This study did not influence or alter the training sessions in any way. Training data collection for this study was carried out at the soccer club's outdoor training pitches. Accumulated total minutes of all training sessions per week are presented in Table 1. Each training session included the warm-up, main phase and slow-down phase plus stretching.

Table 1. Training sessions during the 10-week period.

Weeks (w)	w1	w2	w3	w4	w5	w6	w7	w8	w9	w10
Session duration (total minutes)	385	250	294	285	317	280	274	316	331	270

2.3. Internal Training Load/Intensity Quantification

During training sessions, the CR10-point scale, adapted by Foster et al., was applied [16]. Specifically, thirty minutes after the end of each training session, players rated their RPE value using an app on a tablet. The scores provided by the players were then multiplied by the training duration to obtain the s-RPE [16,37]. The players were previously familiarized with the scale, and all answers were provided individually to avoid non-valid scores.

2.4. Wellness Quantification

Approximately 30 min before each training session, each player was asked to provide the Hooper Index (HI) scores using an app on a tablet. This index includes four categories: fatigue, stress, muscle soreness (scale of 1–7, in which 1 is very, very low and 7 is very, very high), and the quality of sleep of the night that preceded the evaluation (scale of 1–7, in which 1 is very, very bad and 7 is very, very good) [27].

2.5. Calculations of Training Indexes

Through stress, fatigue, DOMS and sleep quality, accumulated data by week were calculated, which includes the summation of each value provided by each training session. Through s-RPE, the following variables were calculated: (i) TM (mean of training load during the seven days of the week divided by the standard deviation of the training load of the seven days) [25,35];

$$\text{TM} = \frac{\text{mean of training load during the seven days of the week}}{\text{standard of training load during the seven days of the week}};$$

(ii) TS (sum of the training loads for all training sessions during a week multiplied by training monotony) [25,35];

$$\text{TS} = \text{sum of the training loads for all training sessions during a week} * \text{TM};$$

Finally, (iii) ACWR (dividing the acute workload, i.e., the 1-week rolling workload data, by the chronic workload, i.e., the rolling 4-week average workload data) [38–40].

$$\text{ACWR} = \frac{\text{acute workload (most recent week)}}{\text{chronic workload (last 4 weeks)}}.$$

2.6. Statistical Analysis

Data were analyzed using SPSS version 22.0 (SPSS Inc., Chicago, IL, USA) for Windows. Initially, descriptive statistics were used to describe and characterize the sample. The Shapiro–Wilk and Levene tests were used to test the assumption of normality and homoscedasticity, respectively. Then, One Way ANOVA was used with the Bonferroni post hoc test to compare player positions and independent *t*-test was used to compare player status [41]. Pearson product–moment correlation coefficient was calculated between s-RPE and HI scores with the following thresholds: ≤ 0.1, trivial; >0.1–0.3, small; >0.3–0.5, moderate; >0.5–0.7, large; >0.7–0.9, very large; >0.9–1.0, almost perfect. Results were considered significant with $p \leq 0.05$.

3. Results

The weekly changes in DOMS, stress, sleep and fatigue over the 10 week period are presented in Figure 1. Overall, DOMS presented the highest value in week 8 (13.6 arbitrary units (AU)) and the lowest in week 2 (6.8 AU); stress presented the highest value in week 6 (12.0 AU) and the lowest in week 10 (7.3 AU); sleep presented the highest value in week 8 (14.9 AU) and the lowest in week 2 (10.4 AU); and fatigue presented the highest value in week 8 (14.7 AU) and the lowest in week 2 (8.7 AU).

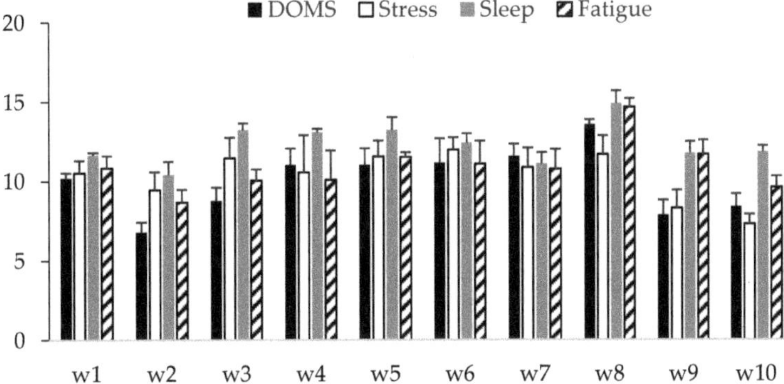

Figure 1. Description of weekly DOMS, stress, sleep and fatigue during the 10 weeks in AU (arbitrary units).

Training monotony and strain over the 10 week period are presented in Figure 2. Overall, training monotony presented the highest value in week 1 (6.3 AU) and the lowest in week 7 (2.7 AU); training strain presented the highest value in week 2 (9665.1 AU) and the lowest in week 7 (3957.6 AU).

ACWR over the 10 week period is presented in Figure 3. Overall, ACWR presented the highest value in week 5 (1.11 AU) and the lowest in week 10 (0.86 AU).

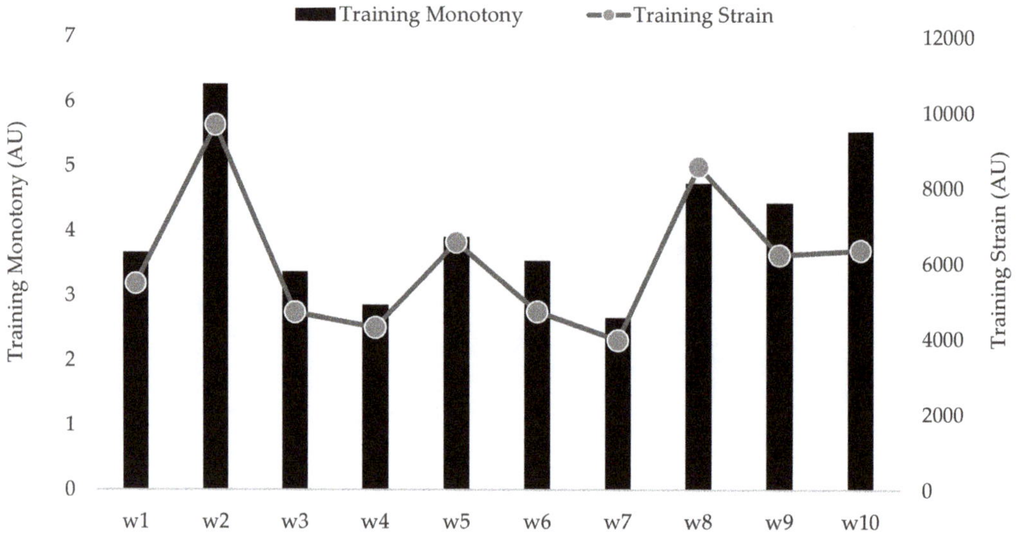

Figure 2. Description of training monotony and training strain during the 10 weeks.

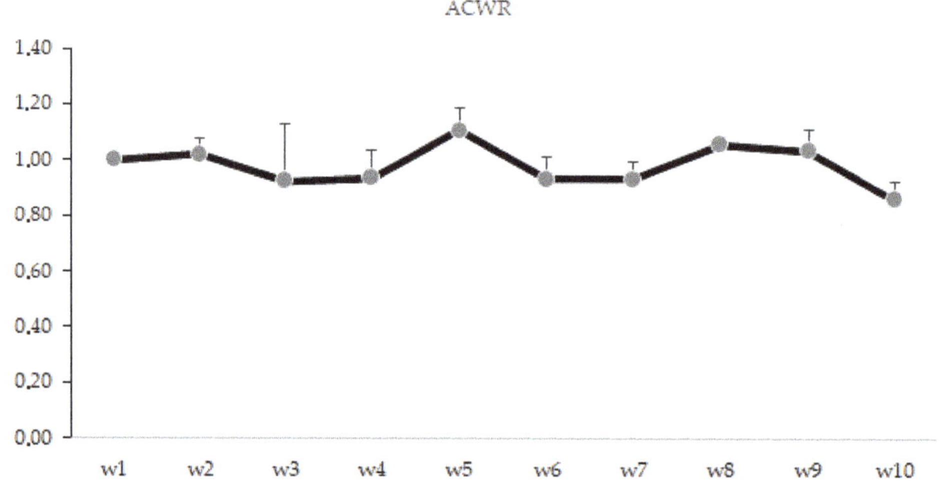

Figure 3. Description of ACWR during the 10 weeks in AU (arbitrary units).

Table 2 presents comparisons for all measures between player positions, while Table 3 presents comparisons between player status. There were no significant differences between player positions nor status.

Table 2. Descriptive statistics (mean ± SD) of weekly stress, fatigue, DOMS, sleep, training monotony, training strain and ACWR between playing positions.

Measures (AU)	Defenders	Central Midfielders	Wide Midfielders	Strikers	F	P
Stress	12.1 ± 2.8	8.9 ± 1.1	10.7 ± 1.2	9.8 ± 2.8	1.830	0.185
Fatigue	11.5 ± 2.2	9.9 ± 0.6	11.4 ± 1.8	10.7 ± 2.4	0.744	0.542
DOMS	9.7 ± 2.2	9.2 ± 2.2	11.2 ± 2.4	9.7 ± 2.7	0.513	0.680
Sleep	12.9 ± 1.5	12.6 ± 0.9	11.9 ± 1.1	12.2 ± 1.3	0.589	0.632
TM	4.3 ± 0.6	4.6 ± 1.0	4.9 ± 0.3	4.4 ± 0.9	0.394	0.759
TS	82.3 ± 9.3	176.4 ± 167.2	125.9 ± 66.4	199.2 ± 258.3	0.518	0.676
ACWR	0.97 ± 0.02	0.97 ± 0.03	0.96 ± 0.01	0.98 ± 0.03	0.240	0.867

Abbreviations: AU, arbitrary units; DOMS, delayed onset muscle soreness; TM, training monotony; TS, training strain; ACWR, acute: chronic workload ratio.

Table 3. Descriptive statistics (mean ± SD) of weekly stress, fatigue, DOMS, sleep, training monotony, training strain and ACWR between playing status.

Measures (AU)	Starters	Non-Starters	T	P
Stress	10.7 ± 2.8	9.8 ± 1.6	0.808	0.430
Fatigue	10.9 ± 2.0	10.8 ± 1.9	0.157	0.877
DOMS	10.5 ± 2.7	9.1 ± 1.2	1.339	0.198
Sleep	12.5 ± 1.4	12.2 ± 0.9	0.482	0.636
TM	4.3 ± 0.7	4.8 ± 0.8	−1.404	0.178
TS	137.5 ± 173.9	160.1 ± 135.2	−0.306	0.763
ACWR	0.96 ± 0.02	0.98 ± 0.03	−1.263	0.224

Abbreviations: AU, arbitrary units; DOMS, delayed onset muscle soreness; TM, training monotony; TS, training strain; ACWR, acute: chronic workload ratio.

Considering the overall team, there were some associations between Hooper Index categories and s-RPE indexes. Specifically, the following correlations were found: stress and fatigue (0.693, $p < 0.01$); stress and DOMS (0.593, $p < 0.01$); stress and TS (−0.516, $p < 0.05$); fatigue and DOMS (0.688, $p < 0.01$). Table 4 presents all correlations.

Table 4. Correlation analysis between measures for overall team.

Measures	β0	β1	β2	β3	β4	β5	β6
Stress (β0)	1.00						
Fatigue (β1)	**0.693**	1.00					
DOMS (β2)	**0.593**	**0.688**	1.00				
Sleep (β3)	0.412	−0.037	−0.001	1.00			
TM (β4)	−0.155	0.165	−0.996	−0.248	1.00		
TS (β5)	**−0.516**	−0.411	−0.329	−0.365	0.043	1.00	
ACWR (β6)	−0.071	0.133	0.148	−0.057	0.012	0.300	1.00

Correlations are highlighted in bold ($p \leq 0.05$). Abbreviations: DOMS, delayed onset muscle soreness; TM, training monotony; TS, training strain; ACWR, acute: chronic workload ratio.

4. Discussion

The aims of this study were: (a) to describe the weekly responses regarding internal training load parameters (TM, TS and ACWR) derived from perceived exertion; (b) to describe weekly variations of wellness markers [stress, fatigue, DOMS and sleep quality]; (c) to identify whether the (unknown) seasonal variations in both load and wellness measures are dependent upon playing position and status; and (d) to analyze possible associations between the several training measures and wellness markers. The main results of the study were: the peak on stress occurred in the week immediately subsequent to when the highest ACWR values were observed, while the lowest values of either stress or ACWR were reported concomitantly in the same week period; best self-reported sleep quality was identified simultaneously in the week where peaks of fatigue/DOMS occurred whilst the inverse also holds true, that is, the smallest indices of DOMS, fatigue and sleep quality temporally coincided; dynamics of perceived load and wellness variations during a 10-week in-season period were similar for starters and reserve players; and playing position was also not a factor affecting load distribution and wellness sensation in women's soccer.

A slight spike in the ACWR of s-RPE (week 5) likely induced players into a greater perception of stress in the nearest next period (week 6). In the same response pattern, ACWR and stress indices reached the lowest magnitudes together in the end of the monitoring period (week 10). It is important to note that, despite ACWR derived from s-RPE not being a good predictor of injury [42], our outcomes observed across the in-season period did not reach the suggested "danger zone" (ACWR $\geq$ 1.5) [43]. This result is in line with another study performed with 65 players from Division I in the United States, which reported this ratio was not associated with injury [44], although some studies reported associations between ACWR and non-contact injury occurrences [42,45–48]. In fact, ACWR score has also been related to perceived effort in women athletes [49]. In addition, TMs across the monitoring weeks were all superior to the traditional cut-point of 2 AU [17]. Although different in magnitude, apparently there were four "valleys" (e.g., week 3 to 4 and week 9 to 10) and three peaks in TM/TS behaviors and this includes week 5 (Figure 2). In this way, women players may be sensitive to simultaneous changes (i.e., some increase) in ACWR, TS and TM—in particular when the first factor notably raises—thus reflecting an ensuing exacerbated stress sensation. As some studies have shown differences in the types of training depending on the moment of the season (i.e., pre, early, mid or end season) [25,50], it is possible link these to the body's adaptations in attempting to accommodate a distinguished training intensity/duration (see Table 1) delivered in this mid-season moment. Owing to the existing relationship between stress symptoms and injury likelihood [51] as well as negative training-induced central adaptations [52], the close monitoring of women players is required at the time of which ACWR and stress spikes happen, aiming to avoid potential time-loss injuries and occasional performance declines.

The present work failed to identify significant relationships between players' sleep quality, wellness perceptions and training indexes, although the dynamics of between-week changes in sleep accompanied the minimum and maximum values of both perceived fatigue and DOMS. Many scientific investigations have confirmed the direct impact of training intensity [53–55] and well-being [46,56] upon sleep measures in soccer. Here, the absence of meaningful correlations of sleep and all other dependent variables may be attributed to a potential lack of sensitivity provided by discrete, self-reported sleep quality metrics. As, for example, sleep questionnaire (subjective) responses do not often match actigraphy-derived or polysomnography (objective) parameters [57]. Nevertheless, when sudden changes are observed in both fatigue and DOMS (especially from week 7 to 8 where TS and TM also increased) attention is required because exacerbated levels of both fatigue and DOMS markers, as obtained by the Hooper Index method, may lead to impaired technical performance [58]. Interestingly, Douchet et al. [59] showed that a week with more accelerations and decelerations induced increased fatigue as observed in the present study by the greater RPE and Hooper index. The authors reported that the objectives of each training session during the week (i.e., technical, tactical, or physical) can contribute to defining the fatigue levels, this being an excellent way to manage the athletes' training load [59].

On the other hand, this is the period in which women soccer players slept better, at least according to their reports. This could be associated with a necessity to restore the systemic homeostasis as illustrated by the elevated tiredness and pain sensations [60].

Therefore, according to the current results, encouraging sleep hygiene strategies could be of particular interest across an intensified training period. Otherwise, the women players choose to extend their sleep opportunities to improve the recovery process and prevent occasions of poor performance in training sessions and official matches due to the high levels of fatigue and muscle soreness. Of note, when the training duration was shortened (e.g., weeks 2 and 7), TM and TS were also impacted in the same direction of declines, thus suggesting that it could be a reasonable strategy when it is necessary to adequately recover players. Notwithstanding, TM and TS peaked respectively in weeks 1 and 2, which is similar to research into male soccer [24,36] and might illustrate the challenge of female athletes to deal with training demands imposed at the beginning of a season.

A key finding of the present investigation was that variations of selected internal load parameters and wellness responses were independent of playing status. Positional role also had a minimal influence on the in-season fluctuations discussed above. One possible explanation for these results could be associated with very small subsample sizes of four to five players, which was not enough to demonstrate significant differences, as demonstrated by previous studies [25,61]. Locomotor capacity and the recovery states of women soccer players are sensitive to the accumulated load demands [62] implying that players' rotation might be needed to counteract such worst-case conditions (e.g., when training was supposedly intensified and/or extended; weeks 5 and 8). In this sense and despite there being players who were selected as starters more often, there should be changes in the coaches' choices depending on these intensified periods. This approach will help to reduce the differences between the playing status in the squad. Therefore, and despite there being players who are the preferred starters, regular changes in the main squad may occur as a function of such intensified periods, which helps to justify the lack of differences according to statuses. While the absence of differences in the analysis of starters versus reserve players is in agreement with the current evidence [24,50], the playing position had a significant influence in an almost matched male sample considering Hooper's Index measures [30]. In this sense, it is necessary to point out the possibility that, in women's soccer, the traditional position-related outputs might be more limited or even non-existent. To be explicit, fitness testing data of past studies indicated that aerobic power was similar across outfield playing positions in Norwegian elite women players [63]. Furthermore, research on college players from Division I in the United States demonstrated that lower limb power, change-of-direction [64], agility, speed and acceleration qualities [65] did not differ according to positional role. Recently, it was confirmed that some of these tests may reflect the in-game running outputs in first Division league Portuguese women soccer players [31], thus suggesting that a systematic discrepancy in game demands depending on player position would not be so evident in females, which could be observed in the study of Vescovi et al. [32], as compared to substantial between-position differences generally found in male counterparts [66,67]. More specific to the present context, external training load in a first Division Spanish women's team revealed no significant differences as a function of playing position (central defenders, wide defenders, central midfielders, wide midfielders and strikers) [68]. Results from a meta-analysis confirm the internal–external training load associations in team sports [10], thereby again making our result concerning training outcomes being independent of playing position compatible with the literature. In sum, it seems that modern soccer demands cause adjustments by the coaching staff in order to deliver relatively equalized training stimuli to women players regardless of their positional role and whether beginning the matches in the starting line-up or on the bench.

Finally, an inverse relationship existed between stress and RPE in the monitored population of women elite soccer players. It was the only significant correlation identified between s-RPE and Hooper Index categories, which contradicts two previous studies in male professional soccer players reporting a number of direct associations amongst RPE and Hooper Index indices [33,69]. However, the large negative s-RPE-stress correlation found here is something that provides preliminary evidence of the possible harmful consequences of higher stress levels on the perception of degree of efforts expended in practice sessions.

This inverse relationship seems to be in line with a recent study that reported higher values of RPE and Hooper index categories with the exception of stress during a higher intensity week [59]. This is in accordance with the notion of Paul et al. [70], where it cannot be discarded that high stress levels may promote a suboptimal psychological interaction with the question being asked relating to perceived effort. As a result, when collecting s-RPE in the context of women's soccer, it is important to consider the prominent interference of stress values that could potentially underestimate athletes' perceived training load demands.

Our present study provides coaches and technical staff with recent knowledge about the weekly variations of TM, TS and ACWR through s-RPE, and the Hooper index categories, in order to have all players available for competition. In addition, this study allows coaches to understand that all these measures vary according to the intensity and duration of the sessions throughout the week. Therefore, planning the structure and periodization of the objectives (i.e., technical, tactical, or physical) and the use of measures of intensity (e.g., RPE) are essential to inducing good adaptations in female athletes.

Limitations of the current study should not be overlooked and includes the fact that: (i) data collections encompassed a single-club, suggesting caution in attempts of extrapolating results to a variety of other teams/leagues; (ii) there was a relatively small sample size when the women athletes were grouped into distinct outfield playing positions; (iii) fatigue levels were measured only via self-reported ratings but not gold standard measures; (iv) non-concomitant consideration for situational factors that can modulate internal load responses during a season; and (v) a lack of match load information when interpreting training outcomes.

5. Conclusions

To conclude, the in-season spike in ACWR calculated through s-RPE may induce women soccer players into experiencing a nearby subsequent peak of perceived stress. Additionally, higher levels of fatigue occur with levels of stress and higher levels of DOMS occur with high levels of stress and fatigue. Reducing training duration could diminish s-RPE-derived strain and monotony indices. Importantly, any in-season variations across a 10-week period concerning internal load and perceived wellness seems independent of position and status in outfield players, although some caution should be taken into consideration. Finally, the higher the players' reported stress, the lower the observed s-RPE, thus possibly indicating a mutual interference of experienced stress levels on the assimilation of training loads by women professional soccer players.

Author Contributions: Conceptualization, R.F. and R.O.; methodology, R.F., J.P.B., H.N., F.M.C. and R.O.; software, R.F., H.N. and R.O.; validation, V.M.R., R.O. and J.P.B.; formal analysis, R.F., R.O. and A.D.M.; investigation, R.O., R.F., A.D.M., J.P.B. and F.M.C.; resources, R.O., R.F., A.D.M., J.P.B., V.M.R. and F.M.C.; data curation, R.F., R.O. and H.N.; writing—original draft preparation, R.O., J.P.B., L.H.P.V., F.M.C. and R.F.; writing—review and editing, R.O., L.H.P.V., R.F., A.D.M., J.P.B. and F.M.C.; visualization, R.O., R.F., J.P.B. and F.M.C.; supervision, R.O. and F.M.C.; project administration, R.O., J.P.B. and R.F.; funding acquisition, R.O. and J.P.B. All authors have read and agreed to the published version of the manuscript.

Funding: This research was funded by the Portuguese Foundation for Science and Technology, I.P., Grant/Award Number UIDP/04748/2020. Victor Machado Reis was funded by FCT–Fundação para a Ciência e Tecnologia (UID04045/2020). This work was also funded by Fundação para a Ciência e Tecnologia/ Ministério da Ciência, Tecnologia e Ensino Superior through national funds and when applicable co-funded EU funds under the project UIDB/50008/2020.

Institutional Review Board Statement: The study was conducted according to the guidelines of the Declaration of Helsinki and approved by the Ethics Committee of Polytechnic Institute of Santarém (252020Desporto).

Informed Consent Statement: Written informed consent was obtained from the participants to publish this paper.

Data Availability Statement: The data presented in this study are available on request from the corresponding author.

Acknowledgments: The authors would like to thank the team's coaches and players for their cooperation during all data collection procedures. Luiz H Palucci Vieira: ongoing PhD fellowship provided by São Paulo Research Foundation—FAPESP, under process number [#2018/02965-7]. "The opinions, hypotheses and conclusions or recommendations expressed in this material are the responsibility of the authors and do not necessarily reflect the views of FAPESP".

Conflicts of Interest: The authors declare no conflict of interest. The funders had no role in the design of the study; in the collection, analyses, or interpretation of data; in the writing of the manuscript, or in the decision to publish the results.

References

1. Miguel, M.; Oliveira, R.; Loureiro, N.; García-Rubio, J.; Ibáñez, S.J. Load measures in training/match monitoring in soccer: A systematic review. *Int. J. Environ. Res. Public Health* **2021**, *18*, 2721. [CrossRef]
2. Impellizzeri, F.M.; Marcora, S.M.; Coutts, A.J. Internal and External Training Load: 15 Years On. *Int. J. Sports Physiol. Perform.* **2019**, *14*, 270–273. [CrossRef]
3. Gabbett, T.J.; Nassis, G.P.; Oetter, E.; Pretorius, J.; Johnston, N.; Medina, D.; Rodas, G.; Myslinski, T.; Howells, D.; Beard, A.; et al. The athlete monitoring cycle: A practical guide to interpreting and applying training monitoring data. *Br. J. Sports Med.* **2017**, *51*, 1451–1452. [CrossRef]
4. Weaving, D.; Beggs, C.; Dalton-Barron, N.; Jones, B.; Abt, G. Visualizing the complexity of the athlete-monitoring cycle through principal-component analysis. *Int. J. Sports Physiol. Perform.* **2019**, *14*, 1304–1310. [CrossRef]
5. Oliveira, R.; Brito, J.P.; Martins, A.; Mendes, B.; Marinho, D.A.; Ferraz, R.; Marques, M.C. In-season internal and external training load quantification of an elite European soccer team. *PLoS ONE* **2019**, *14*, e0209393. [CrossRef]
6. Ball, S.; Halaki, M.; Orr, R. Movement Demands of Rugby Sevens in Men and Women: A Systematic Review and Meta-Analysis. *J. Strength Cond. Res.* **2019**, *33*, 3475–3490. [CrossRef]
7. Sánchez-Sáez, J.A.; Sánchez-Sánchez, J.; Martínez-Rodríguez, A.; Felipe, J.L.; García-Unanue, J.; Lara-Cobos, D. Global Positioning System Analysis of Physical Demands in Elite Women's Beach Handball Players in an Official Spanish Championship. *Sensors* **2021**, *21*, 850. [CrossRef]
8. Clarke, A.C.; Anson, J.; Pyne, D. Physiologically based GPS speed zones for evaluating running demands in Women's Rugby Sevens. *J. Sports Sci.* **2015**, *33*, 1101–1108. [CrossRef]
9. Bradley, P.S.; Vescovi, J.D. Velocity Thresholds for Women's Soccer Matches: Sex Specificity Dictates High-Speed-Running and Sprinting Thresholds—Female Athletes in Motion (FAiM). *Int. J. Sports Physiol. Perform.* **2015**, *10*, 112–116. [CrossRef]
10. McLaren, S.J.; Macpherson, T.W.; Coutts, A.J.; Hurst, C.; Spears, I.R.; Weston, M. The Relationships Between Internal and External Measures of Training Load and Intensity in Team Sports: A Meta-Analysis. *Sport. Med.* **2018**, *48*, 641–658. [CrossRef]
11. Nobari, H.; Alves, A.R.; Haghighi, H.; Clemente, F.M.; Carlos-Vivas, J.; Pérez-Gómez, J.; Ardigò, L.P. Association between training load and well-being measures in young soccer players during a season. *Int. J. Environ. Res. Public Health* **2021**, *18*, 4451. [CrossRef] [PubMed]
12. Clemente, F.M.; Clark, C.; Castillo, D.; Sarmento, H.; Nikolaidis, P.T.; Rosemann, T.; Knechtle, B. Variations of training load, monotony, and strain and dose-response relationships with maximal aerobic speed, maximal oxygen uptake, and isokinetic strength in professional soccer players. *PLoS ONE* **2019**, *14*, e0225522. [CrossRef]
13. Rabbani, A.; Kargarfard, M.; Castagna, C.; Clemente, F.M.; Twist, C. Associations between Selected Training Stress Measures and Fitness Changes in Male Soccer Players. *Int. J. Sports Physiol. Perform.* **2019**, *14*, 1050–1057. [CrossRef]
14. Impellizzeri, F.M.; Rampinini, E.; Coutts, A.J.; Sassi, A.; Marcora, S.M. Use of RPE-based training load in soccer. *Med. Sci. Sports Exerc.* **2004**, *36*, 1042–1047. [CrossRef]
15. Fanchini, M.; Ferraresi, I.; Petruolo, A.; Azzalin, A.; Ghielmetti, R.; Schena, F.; Impellizzeri, F.M. Is a retrospective RPE appropriate in soccer? Response shift and recall bias. *Sci. Med. Footb.* **2017**, *1*, 53–59. [CrossRef]
16. Foster, C.; Florhaug, J.A.; Franklin, J.; Gottschall, L.; Hrovatin, L.A.; Parker, S.; Doleshal, P.; Dodge, C. A New Approach to Monitoring Exercise Training. *J. Strength Cond. Res.* **2001**, *15*, 109–115. [CrossRef]
17. Foster, C. Monitoring training in athletes with reference to overtraining syndrome. *Med. Sci. Sports Exerc.* **1998**, *30*, 1164–1168. [CrossRef] [PubMed]
18. Murray, N.B.; Gabbett, T.J.; Townshend, A.D.; Blanch, P. Calculating acute: Chronic workload ratios using exponentially weighted moving averages provides a more sensitive indicator of injury likelihood than rolling averages. *Br. J. Sports Med.* **2017**, *51*, 749–754. [CrossRef]
19. Maupin, D.; Schram, B.; Canetti, E.; Orr, R. The Relationship Between Acute: Chronic Workload Ratios and Injury Risk in Sports: A Systematic Review. *Open Access J. Sport. Med.* **2020**, *11*, 51–75. [CrossRef]
20. Impellizzeri, F.M.; Ward, P.; Coutts, A.J.; Bornn, L.; Mccall, A. Training load and injury: Part 1. The devil is in the detail—challenges to applying the current research in the training load and injury field. *J. Orthop. Sport. Phys. Ther.* **2020**, *50*, 574–576. [CrossRef] [PubMed]

21. Impellizzeri, F.M.; Ward, P.; Coutts, A.J.; Mccall, A. Training load and injury part 2: Questionable research practices hijack the truth and mislead well-intentioned clinicians. *J. Orthop. Sport. Phys. Ther.* **2020**, *50*, 577–584. [CrossRef]
22. Impellizzeri, F.M.; Tenan, M.S.; Kempton, T.; Novak, A.; Coutts, A.J. Acute: Chronic workload ratio: Conceptual issues and fundamental pitfalls. *Int. J. Sports Physiol. Perform.* **2020**, *15*, 907–913. [CrossRef]
23. Fernandes, R.; Oliveira, R.; Martins, A.; Paulo Brito, J. Internal training and match load quantification of one-match week schedules in female first league Portugal soccer. *Cuad. Psicol. del Deport.* **2021**, *21*, 126–138. [CrossRef]
24. Oliveira, R.; Vieira, L.H.P.; Martins, A.; Brito, J.P.; Nalha, M.; Mendes, B.; Clemente, F.M. In-season internal and external workload variations between starters and non-starters—a case study of a top elite european soccer team. *Medicina* **2021**, *57*, 645. [CrossRef]
25. Clemente, F.M.; Silva, R.; Castillo, D.; Arcos, A.L.; Mendes, B.; Afonso, J. Weekly load variations of distance-based variables in professional soccer players: A full-season study. *Int. J. Environ. Res. Public Health* **2020**, *17*, 3300. [CrossRef]
26. Duignan, C.; Doherty, C.; Caulfield, B.; Blake, C. Single-item self-report measures of team-sport athlete wellbeing and their relationship with training load: A systematic review. *J. Athl. Train.* **2020**, *55*, 944–953. [CrossRef]
27. Hooper, S.L.; Mackinnon, L.T. Monitoring Overtraining in Athletes: Recommendations. *Sport. Med.* **1995**, *20*, 321–327. [CrossRef]
28. McLean, B.D.; Coutts, A.J.; Kelly, V.; McGuigan, M.R.; Cormack, S.J. Neuromuscular, endocrine, and perceptual fatigue responses during different length between-match microcycles in professional rugby league players. *Int. J. Sports Physiol. Perform.* **2010**, *5*, 367–383. [CrossRef]
29. Fessi, M.S.; Nouira, S.; Dellal, A.; Owen, A.; Elloumi, M.; Moalla, W. Changes of the psychophysical state and feeling of wellness of professional soccer players during pre-season and in-season periods. *Res. Sport. Med.* **2016**, *24*, 375–386. [CrossRef]
30. Clemente, F.M.; Mendes, B.; Nikolaidis, P.T.; Calvete, F.; Carriço, S.; Owen, A.L. Internal training load and its longitudinal relationship with seasonal player wellness in elite professional soccer. *Physiol. Behav.* **2017**, *179*, 262–267. [CrossRef]
31. Gonçalves, L.; Clemente, F.M.; Barrera, J.I.; Sarmento, H.; González-Fernández, F.T.; Palucci Vieira, L.H.; Figueiredo, A.J.; Clark, C.C.T.; Carral, J.M.C. Relationships between Fitness Status and Match Running Performance in Adult Women Soccer Players: A Cohort Study. *Medicina* **2021**, *57*, 617. [CrossRef]
32. Vescovi, J.D.; Favero, T.G. Motion characteristics of women's college soccer matches: Female athletes in motion (faim) study. *Int. J. Sports Physiol. Perform.* **2014**, *9*, 405–414. [CrossRef]
33. Moalla, W.; Fessi, M.S.; Farhat, F.; Nouira, S.; Wong, D.P.; Dupont, G. Relationship between daily training load and psychometric status of professional soccer players. *Res. Sport. Med.* **2016**, *24*, 387–394. [CrossRef]
34. Nobari, H.; Oliveira, R.; Clemente, F.M.; Adsuar, J.C.; Pérez-Gómez, J.; Carlos-Vivas, J.; Brito, J.P. Comparisons of accelerometer variables training monotony and strain of starters and non-starters: A full-season study in professional soccer players. *Int. J. Environ. Res. Public Health* **2020**, *17*, 6547. [CrossRef] [PubMed]
35. Nobari, H.; Praça, G.M.; Clemente, F.M.; Pérez-Gómez, J.; Carlos Vivas, J.; Ahmadi, M. Comparisons of new body load and metabolic power average workload indices between starters and non-starters: A full-season study in professional soccer players. *Proc. Inst. Mech. Eng. Part P J. Sport. Eng. Technol.* **2020**, *235*, 105–113. [CrossRef]
36. Oliveira, R.; Martins, A.; Nobari, H.; Nalha, M.; Mendes, B.; Clemente, F.M.; Brito, J.P. In—season monotony, strain and acute/chronic workload of perceived exertion, global positioning system running based variables between player positions of a top elite soccer team. *BMC Sports Sci. Med. Rehabil.* **2021**, *13*, 1–10. [CrossRef]
37. Foster, C.; Hector, L.L.; Welsh, R.; Schrager, M.; Green, M.A.; Snyder, A.C. Effects of specific versus cross-training on running performance. *Eur. J. Appl. Physiol. Occup. Physiol.* **1995**, *70*, 367–372. [CrossRef]
38. Dalen-Lorentsen, T.; Bjørneboe, J.; Clarsen, B.; Vagle, M.; Fagerland, M.W.; Andersen, T.E. Does load management using the acute:chronic workload ratio prevent health problems? A cluster randomised trial of 482 elite youth footballers of both sexes. *Br. J. Sports Med.* **2021**, *55*, 108–114. [CrossRef]
39. Myers, N.L.; Aguilar, K.V.; Mexicano, G.; Farnsworth, J.L.; Knudson, D.; Kibler, W.B.E.N. The Acute:Chronic Workload Ratio Is Associated with Injury in Junior Tennis Players. *Med. Sci. Sports Exerc.* **2020**, *52*, 1196–1200. [CrossRef]
40. Impellizzeri, F.; Woodcock, S.; Coutts, A.J.; Fanchini, M.; McCall, A.; Vigotsky, A.D. Acute to random chronic workload ratio is 'as' associated with injury as acute to actual chronic workload ratio: Time to dismiss ACWR and its components. *SportRχiv* **2020**. [CrossRef]
41. Buchheit, M.; Mendez-Villanueva, A. Changes in repeated-sprint performance in relation to change in locomotor profile in highly-trained young soccer players. *J. Sports Sci.* **2014**, *32*, 1309–1317. [CrossRef] [PubMed]
42. Fanchini, M.; Rampinini, E.; Riggio, M.; Coutts, A.J.; Pecci, C.; Mccall, A. Despite association, the acute: Chronic work load ratio does not predict non-contact injury in elite footballers. *Sci. Med. Footb.* **2018**, *2*, 108–114. [CrossRef]
43. Gabbett, T.J. The training—injury prevention paradox: Should athletes be training smarter and harder? *Br. J. Sports Med.* **2016**, *50*, 273–280. [CrossRef]
44. Xiao, M.; Nguyen, J.N.; Hwang, C.E.; Abrams, G.D. Increased Lower Extremity Injury Risk Associated With Player Load and Distance in Collegiate Women's Soccer. *Orthop. J. Sport. Med.* **2021**, *9*, 1–8. [CrossRef]
45. Bowen, L.; Gross, A.S.; Gimpel, M.; Bruce-Low, S.; Li, F.-X. Spikes in acute:chronic workload ratio (ACWR) associated with a 5–7 times greater injury rate in English Premier League football players: A comprehensive 3-year study. *Br. J. Sports Med.* **2020**, *54*, 731–738. [CrossRef]
46. Watson, A.; Brickson, S.; Brooks, A.; Dunn, W. Subjective well-being and training load predict in-season injury and illness risk in female youth soccer players. *Br. J. Sports Med.* **2017**, *51*, 194–199. [CrossRef] [PubMed]

47. Curtis, R.M.; Huggins, R.A.; Benjamin, C.L.; Sekiguchi, Y.; Lepley, L.K.; Huedo-Medina, T.B.; Casa, D.J. Factors Associated With Noncontact Injury in Collegiate Soccer: A 12-Team Prospective Study of NCAA Division I Men's and Women's Soccer. *Am. J. Sport. Med.* **2021**, *49*, 3076–3087. [CrossRef] [PubMed]

48. Ishida, A.; Beaumont, J.S. Difference in Internal and External Workloads between Non-Injured and Injured Groups in Collegiate Female Soccer Players. *Int. J. Kinesiol. Sport. Sci.* **2020**, *8*, 26–32. [CrossRef]

49. Foster, I.; Byrne, P.J.; Moody, J.A.; Fitzpatrick, P.A. Monitoring Training Load Using the Acute: Chronic Workload Ratio in Non-Elite Intercollegiate Female Athletes. *ARC J. Res. Sport. Med.* **2018**, *3*, 22–28.

50. Martins, A.D.; Oliveira, R.; Brito, J.P.; Loureiro, N.; Querido, S.M.; Nobari, H. Intra-season variations in workload parameters in europe's elite young soccer players: A comparative pilot study between starters and non-starters. *Healthcare* **2021**, *9*, 977. [CrossRef] [PubMed]

51. Clement, D.; Ivarsson, A.; Tranaeus, U.; Johnson, U.; Stenling, A. Investigating the influence of intra-individual changes in perceived stress symptoms on injury risk in soccer. *Scand. J. Med. Sci. Sport.* **2018**, *28*, 1461–1466. [CrossRef]

52. Nédélec, M.; Halson, S.; Abaidia, A.; Ahmaidi, S.; Ne, M.; Dupont, G. Stress, Sleep and Recovery in Elite Soccer: A Critical Review of the Literature. *Sport. Med.* **2015**, *45*, 1387–1400. [CrossRef]

53. Costa, J.A.; Figueiredo, P.; Nakamura, F.Y.; Rebelo, A.; Brito, J. Monitoring Individual Sleep and Nocturnal Heart Rate Variability Indices: The Impact of Training and Match Schedule and Load in High-Level Female Soccer Players. *Front. Physiol.* **2021**, *12*. [CrossRef] [PubMed]

54. Clemente, F.M.; Afonso, J.; Costa, J.; Oliveira, R.; Pino-Ortega, J.; Rico-González, M. Relationships between Sleep, Athletic and Match Performance, Training Load, and Injuries: A Systematic Review of Soccer Players. *Healthcare* **2021**, *9*, 808. [CrossRef] [PubMed]

55. Figueiredo, P.; Costa, J.; Lastella, M.; Morais, J.; Brito, J. Sleep Indices and Cardiac Autonomic Activity Responses during an International Tournament in a Youth National Soccer Team. *Int. J. Environ. Res. Public Health* **2021**, *18*, 2076. [CrossRef]

56. Watson, A.; Brickson, S. Impaired Sleep Mediates the Negative Effects of Training Load on Subjective Well-Being in Female Youth Athletes. *Sport. Health A Multidiscip. Approach* **2018**, *10*, 244–249. [CrossRef] [PubMed]

57. Matthews, K.A.; Patel, S.R.; Pantesco, E.J.; Buysse, D.J.; Kamarck, T.W.; Lee, L.; Hall, M.H. Similarities and Differences in Estimates of Sleep Duration by Polysomnography, Actigraphy, Diary, and Self-reported Habitual Sleep in a Community Sample. *Sleep Health* **2018**, *4*, 96–103. [CrossRef]

58. Selmi, O.; Gonçalves, B.; Ouergui, I.; Levitt, D.E.; Sampaio, J.; Bouassida, A. Influence of Well-Being Indices and Recovery State on the Technical and Physiological Aspects of Play During Small-Sided Games. *J. Strength Cond. Res.* **2021**, *35*, 2802–2809. [CrossRef]

59. Douchet, T.; Humbertclaude, A.; Cometti, C.; Paizis, C.; Babault, N. Quantifying accelerations and decelerations in elite women soccer players during regular in-season training as an index of training load. *Sports* **2021**, *9*, 109. [CrossRef] [PubMed]

60. Dupuy, O.; Douzi, W.; Theurot, D.; Bosquet, L.; Dugué, B. An evidence-based approach for choosing post-exercise recovery techniques to reduce markers of muscle damage, Soreness, fatigue, and inflammation: A systematic review with meta-analysis. *Front. Physiol.* **2018**, *9*, 403. [CrossRef] [PubMed]

61. Clemente, F.M.; Silva, R.; Ramirez-Campillo, R.; Afonso, J.; Mendes, B.; Chen, Y.S. Accelerometry-based variables in professional soccer players: Comparisons between periods of the season and playing positions. *Biol. Sport* **2020**, *37*, 389–403. [CrossRef] [PubMed]

62. Williams, J.H.; Jaskowak, D.J. The effects of match congestion on gait complexity in female collegiate soccer players. *Transl. Sports Med.* **2019**, *2*, 130–137. [CrossRef]

63. Haugen, T.A.; Tønnessen, E.; Hem, E.; Leirstein, S.; Seiler, S. VO2max Characteristics of Elite Female Soccer Players, 1989–2007. *Int. J. Sports Physiol. Perform.* **2014**, *9*, 515–521. [CrossRef]

64. Lockie, R.G.; Moreno, M.R.; Lazar, A.; Orjalo, A.J.; Giuliano, D.V.; Risso, F.G.; Davis, D.L.; Crelling, J.B.; Lockwood, J.R.; Jalilvand, F. The Physical and Athletic Performance Characteristics of Division I Collegiate Female Soccer Players by Position. *J. Strength Cond. Res.* **2018**, *32*, 334–343. [CrossRef]

65. Vescovi, J.D.; Brown, T.D.; Murray, T.M. Positional characteristics of physical performance in Division I college female soccer players. *J. Sport. Med. Phys. Fit.* **2006**, *46*, 221–226.

66. Henrique, L.; Vieira, P.; Carling, C.; Augusto, F.; Rodrigo, B.; Roberto, P.; Santiago, P. Match Running Performance in Young Soccer Players: A Systematic Review. *Sport. Med.* **2019**, *49*, 289–318. [CrossRef]

67. Sarmento, H.; Marcelino, R.; Anguera, M.T.; Campaniço, J.; Matos, N. Match analysis in football: A systematic review. *J. Sports Sci.* **2014**, *32*, 1831–1843. [CrossRef]

68. Romero-Moraleda, B.; Nedergaard, N.J.; Morencos, E.; Ramirez-campillo, R.; Vanrenterghem, J.; Nedergaard, N.J.; Morencos, E. External and internal loads during the competitive season in professional female soccer players according to their playing position: Differences between training and competition. *Res. Sport. Med.* **2021**, *29*, 1–13. [CrossRef] [PubMed]

69. Selmi, O.; Ouergui, I.; Castellano, J.; Levitt, D.; Bouassida, A. Effect of an intensified training period on well-being indices, recovery and psychological aspects in professional soccer players. *Rev. Eur. Psychol. Appl.* **2020**, *70*, 100603. [CrossRef]

70. Paul, D.; Read, P.; Farooq, A.; Jones, L. Factors Influencing the Association Between Coach and Athlete Rating of Exertion: A Systematic Review and Meta- analysis. *Sport. Med.* **2021**, *7*, 1–20. [CrossRef]

International Journal of
Environmental Research and Public Health

MDPI

Article

Does the Number of Substitutions Used during the Matches Affect the Recovery Status and the Physical and Technical Performance of Elite Women's Soccer?

Ronaldo Kobal [1,2,*], Rodrigo Aquino [3,*], Leonardo Carvalho [1], Adriano Serra [1], Rafaela Sander [1], Natan Gomes [1], Vinicius Concon [1], Guilherme Passos Ramos [4,5] and Renato Barroso [1]

1   School of Physical Education, University of Campinas, Campinas 13083-851, Brazil
2   Sport Club Corinthians Paulista, São Paulo 03087-000, Brazil
3   LabSport, Department of Sports, Centre of Physical Education and Sports, Federal University of Espírito Santo, Vitória 29075-910, Brazil
4   Brazilian Football Confederation, Rio de Janeiro 22775-055, Brazil
5   Laboratory of Endocrinology and Metabolism, Federal University of Minas Gerais, Belo Horizonte 31270-901, Brazil
*   Correspondence: rk.sportperformance@gmail.com (R.K.); aquino.rlq@gmail.com (R.A.)

Abstract: The aim of this study was to compare the effect of a new rule for substitutions (four and five) with the rule before the COVID-19 pandemic (up to three) on recovery status, physical and technical performance, internal workload, and recovery process in elite women soccer players. Thirty-eight matches from 2019 to 2020 from the Brazilian Championships were analyzed. All data for the two conditions (≤ 3 and 4–5 substitutions) were compared using an independent t-test. The physical demands measured by a global positioning system (GPS) and the technical (obtained from Instat) and internal workload (rating of perceived exertion [RPE]) were assessed. The recovery process was measured by the total quality recovery (TQR) 24 h after each match. No differences were observed in any physical and technical parameters between 4–5 and ≤ 3 substitutions ($p > 0.05$). Moreover, 4–5 substitutions demonstrated lower RPE ($p < 0.001$) and workload-RPE ($p < 0.001$), higher TQR ($p = 0.008$), and lower time played by the player ($p < 0.001$), compared to ≤ 3. Thus, the new provisory rule for substitutions improved the balance between stress and recovery.

Keywords: contextual factor; football; team sports; female players

Citation: Kobal, R.; Aquino, R.; Carvalho, L.; Serra, A.; Sander, R.; Gomes, N.; Concon, V.; Ramos, G.P.; Barroso, R. Does the Number of Substitutions Used during the Matches Affect the Recovery Status and the Physical and Technical Performance of Elite Women's Soccer? *Int. J. Environ. Res. Public Health* **2022**, *19*, 11541. https://doi.org/10.3390/ijerph191811541

Academic Editor: Paul B. Tchounwou

Received: 15 August 2022
Accepted: 12 September 2022
Published: 14 September 2022

Publisher's Note: MDPI stays neutral with regard to jurisdictional claims in published maps and institutional affiliations.

1. Introduction

Soccer is an intermittent and high-intensity team sport, which depends on endurance, strength, power, and speed abilities [1–3]. Elite women soccer players cover ~10 km, with ~1500 m in high intensity running, and ~200 m in sprint distance during a soccer match [4–6]. In addition, physical performance declines during the matches [3,7], for instance, Bradley et al. [8] observed in the women's UEFA Champions League large decrements in the second half compared to the first half in the distance covered at speeds above 12 km·h^{-1} (1651 vs. 1500 m in the first and second half, respectively). Additionally, Mohr et al. [3] reported that the total distance covered in the first half was longer than the second half (5.28 vs. 5.00 km), and the same trend was also observed for the high intensity running (0.91 vs. 0.70 km), and sprint distance (0.25 vs. 0.21 km). Interestingly, these impairments in physical performance were not observed in the indicators of technical performance, such as successful passes, lost balls, and duels won. Thus, declines in physical performance of soccer players do not seem to affect the technical performance.

Coaches are allowed to use three substitutions per match in an attempt to create a novel team tactical strategy, to replace a player who has become injured, to maintain the physical performance during the match, and to minimize performance decrements in the second half. Bradley et al. [9] compared the physical demand covered during the soccer matches among players who completed the entire match, replaced players (who were substituted), and substituted players (who entered during the match) in the English Premier League. Substitute players covered a higher total distance and high intensity running distance when normalized by the time they played (minutes per minute, m/min) than players who were replaced or completed a full match, and replaced players covered a higher distance than players that completed a full match (120, 116, and 112 m/min; and 9.0, 8.2, and 7.1 m/min, respectively) [9]. Padron-Cabo et al. [10] demonstrated a similar finding in the Spanish League during the season 2014–2015. Starting players who completed matches covered a lower total distance/minute and high intensity running distance/minute than replaced players and substitute players (106, 111, and 111 m/min and 5.1, 6.1, and 6.7 m/min, respectively), with no difference between replaced and substitute.

In general, the measure of distance/minute is used by researchers and practitioners to investigate topics that can influence match running performance [9,11,12]. However, normalized distances should be used with caution because this measure is time-played dependent and did not support decrement of physical performance. For instance, substitute players that played 10 min and covered 120 m/min covered a total distance of the 1200 m, and players who covered 112 m/min for 90 min covered 10,080 m. Thus, relative metrics (i.e., m/min) do not allow for understanding the impact of the whole team performance. Alternatively, to investigate the influence of the substitute players on the whole team performance, it is required to sum metrics from all players who participated in the match.

Recently, the *Fédération Internationale de Football Association* (FIFA) provisionally changed the rules regarding the number of substitutions, allowing up to five by team. The rationale for this change lies in the congested periods of matches that would be experienced by players returning to play following COVID-19 lockdown. The large number of matches in a shorter period of time would cause fatigue to accumulate and increase the risk of injury [13,14]. More substitutions reduce the overall player's match time exposure due to greater players' rotation, which diminishes external and internal load [i.e., total distance covered and rating of perceived exertion (RPE), respectively] in some of the players. In addition, it seems that there is a negative correlation between RPE and next day recovery status, suggesting that athletes who display a high perception of exertion are those that demonstrate low recovery indexes [15]. Interestingly, athletes who show higher RPE and worse recovery status were those that presented higher injury rates and were sick more frequently [16]. Thus, increasing the number of substitutions can contribute to hasten players' recovery after matches and to safeguard the athlete's health by avoiding illness and injury.

Nonetheless, the effect of the number of substitutions and the provisory rule is unclear, especially in women soccer players. Therefore, the purpose of this study was to compare the effect of the number of substitutions on recovery status and the physical and technical performance in elite women soccer players. We hypothesized that the higher number of substitutions would decrease physical demands and internal load and improve the recovery status of elite women soccer players, with no changes in technical performance.

2. Materials and Methods

2.1. Study Design

This retrospective study was conducted during the 2019 and 2020 seasons in the National Brazilian Championships and 38 official matches were analyzed (19 matches in each championship). In 2019, up to three substitutions per match were allowed, whereas in 2020 the rule allowed up to five substitutions. Matches were classified according to the number of substitutions performed (i.e., 0–5). Then, all matches were grouped into 2–3 (21 matches) and 4–5 substitutions (17 matches). The physical performance was measured by GPS and RPE was assessed after all matches. The recovery process was measured by a TQR questionnaire 24 h after each match. In addition, the technical performance was determined by data provided by Instat® (Instat, Moscow, Russia).

2.2. Participants

Twenty-four professional Brazilian women soccer players (28.0 ± 4.6 years, 163.7 ± 5.2 cm, 58.8 ± 7.7 kg, and body fat 22 ± 8.7%) from the same club participated in this study. All players participated in the National Brazilian Championship, seasons 2019 and 2020 (second and first place, respectively), thus attesting their high level of performance. Six players had already participated in the National Brazilian soccer team. Goalkeepers were excluded from the study. All participants were informed about the procedures, benefits, discomforts, and possible risks of the study and signed a free and informed consent before participation. The University's Research Ethics Committee approved the experimental protocol.

2.3. Total Quality Recovery (TQR)

The TQR scale (6–20) was used to provide a means to measure psychophysiological recovery [17]. The athletes answered the question "How do you feel about your recovery?" using the TQR scale, 24 h after each match. All players were previously familiarized with the scale, and the mean results were utilized for analysis.

2.4. Rating of Perceived Exertion (RPE)

Athletes were familiarized with the CR-10 Borg RPE scale as this was part of their training routine. RPE was assessed with the modified Borg 10-point (0–10) scale, which is widely used in both practical and research settings [18]. This method uses a simple question: "How was your match today?". The answer was provided 30 min after the end of training sessions and official matches, by choosing a descriptor and a number from 0–10. We considered the mean of each match in the analysis. The workload-RPE was determined by multiplying each players' playing time (minutes) in each match by the RPE, as described by Foster [18].

2.5. Match Running Performance

All soccer matches activity profiles were obtained via a GPS-system operating at 10 Hz (GPS-units; Playertek, Catapult Innovations, Melbourne, Australia). GPS devices were fitted to the upper back of players using a harness and the same unit was used by each player in all measures to reduce inter-unit measurement error. Units were turned on 10 min before each match. Thirty-eight official matches were measured using GPS. The activity profiles measured were total distance covered (km), total sprint distance (m) (speed > 18 km·h^{-1}), number of accelerations and decelerations (>3 m·s^{-2}), and top-speed (km·h^{-1}). For the analysis, we used the sum of all GPS metrics of all players that participated in each match, except for the top speed, total distance/minute and total sprint distance/minute that we used the mean of demand of all players that participated by matches.

2.6. Technical Analysis

Soccer specific technical performance indicators were measured, and the following match statistics were used: goals scored, goals conceded, total shots, shots on target, assistance, passes, accurate passes, crosses, lost ball, recovered ball, dribbles, and the Instat® index. Data were obtained from Instat® (Instat®, Moscow, Russia), a private company which provides teams' technical performance assessments worldwide. The Instat index® is calculated based on a unique set of key parameters for each playing position (12–14 performance parameters, depending on the position during the game), with a higher numerical value indicating better performance. The exact calculations are trademarked and known only to the manufacturer of the platform [19]. In most general terms, an automatic algorithm considers the player's contribution to the team's success, the significance of their actions, opponent's level, and the level of the competition they play in. The rating is created automatically, and each parameter has a factor which changes depending on the number of actions and events in the match. The weight of the action factors differs depending on the player's position. The key factors included in the calculation of the Instat index® are position specific and include tackling, aerial duels, set pieces in defense, interceptions for central defenders; number of crosses, number of passes to the penalty area, pressing for full back; playmaking, number of key passes, finishing for central midfielders; pressing, dribbling, finishing, counterattacking for wide midfielders; shooting, finishing, pressing, dribbling for forwards. For the analysis, we used the average of all players in each match.

2.7. Statistical Analysis

Data were visually inspected for the existence of outliers (box-plots), tested for normality (Shapiro-Wilk) and homogeneity (Levene), and are presented as means, standard deviation (SD), and 95% confidence intervals (95%CI) of difference between means. All data of perceived exertion and physical and technical parameters for the 2 conditions (2–3 substitutions and 4–5 substitutions) were analyzed with an independent *t*-test. The magnitudes between condition differences were expressed as standardized mean differences and were interpreted using the following thresholds: <0.2, 0.2–0.6, 0.6–1.2, 1.2–2.0, 2.0–4.0, and >4.0, for trivial, small, moderate, large, very large, and near perfect, respectively [20]. The level of significance was set at $p < 0.05$. Statistical analyses were performed using the software package IBM SPSS (V. 25, SPSS Inc., Chicago, IL, USA).

3. Results

All matches analyzed were classified as follows: eight matches used two substitutions, 13 matches used three substitutions, two matches used four substitutions, and 15 matches used five substitutions. Then, all matches were grouped into 2–3 (21 matches) and 4–5 substitutions (17 matches). There was no match with no or one substitution. Table 1 displays physical and technical match performance, RPE, workload-RPE, TQR values, and time played of players. The condition of 4–5 substitutions demonstrated significantly lower RPE ($p \leq 0.001$), workload-RPE ($p \leq 0.001$), higher TQR ($p = 0.008$), and lower time played by player ($p \leq 0.001$), compared to 2–3 substitutions. No significant differences were observed in the physical and technical parameters between 4–5 and 2–3 substitutions ($p > 0.05$). Figure 1 shows the standardized mean differences between 2–3 and 4–5 substitutions for all variables.

Table 1. Statistics effects between 2–3 and 4–5 substitutions in perceived exertion, physical, and technical variables, mean and SD.

	Variable	2–3 Substitutions	4–5 Substitutions	*p* Value	Mean Difference and 95% CI
Perceived Exertion	RPE (a.u.)	8.1 ± 0.8	6.7 ± 0.7 *	$p < 0.001$	1.4 (0.84–2.00)
	TQR (a.u.)	13.1 ± 0.8	13.9 ± 0.5 *	$p = 0.008$	0.8 (0.2–1.5)
	Time (min)	76.4 ± 4.4	65.5 ± 1.7 *	$p < 0.001$	10.9 (8.6–13.2)
	Workload-RPE (a.u)	623 ± 85	441 ± 53 *	$p < 0.001$	182 (128.7–234.1)
Physical	TD/min (m/min)	106 ± 6	102 ± 6	$p = 0.064$	4 (−0.2–8.1)
	TSD/min (m/min)	7.7 ± 1.2	8.2 ± 1.8	$p = 0.340$	0.5 (−1.5–0.5)
	TD (km)	98.6 ± 5.0	98.0 ± 4.0	$p = 0.650$	0.6 (−2.4–3.8)
	TSD (m)	7002 ± 1191	7465 ± 1614	$p = 0.316$	463 (−1386–460)
	Top Speed (km/h)	26.4 ± 0.6	26.8 ± 0.5	$p = 0.074$	0.4 (−0.8–0.04)
	ACC (n)	715 ± 81	736 ± 69	$p = 0.404$	21 (−71.7–29)
	DECC (n)	913 ± 88	910 ± 67	$p = 0.909$	3 (−50–59)
Technical	Goals conceded (n)	0.38 ± 0.6	0.35 ± 0.6	$p = 0.886$	0.03 (−0.36–0.42)
	Goals scored (n)	2.7 ± 2.2	2.9 ± 1.9	$p = 0.807$	0.2 (−1.5–1.2)
	Assistance (n)	1.8 ± 1.6	1.9 ± 1.5	$p = 0.9$	0.1 (−1.2–1.1)
	Shots (n)	17.9 ± 8	20.8 ± 7	$p = 0.296$	2.9 (−8.6–2.7)
	Shots on goal (n)	7.9± 3.5	9.0 ± 3.1	$p = 0.367$	1.1 (−3.6–1.4)
	Passes (n)	514 ± 94	533 ± 82	$p = 0.537$	19 (−81.7–43.4)
	Crosses (n)	14.0 ± 5.0	16.4 ± 9.0	$p = 0.372$	2.4 (−7.8–3.0)
	Lost balls (n)	75.6 ± 7.0	70.1 ± 9.0	$p = 0.073$	5.5 (−0.5–11.4)
	Recovered balls (n)	58.1 ± 5.0	54.8 ± 6.0	$p = 0.142$	3.3 (−1.1–7.6)
	Dribbles (n)	28.1 ± 9.0	29.2 ± 9.0	$p = 0.734$	1.1 (−7.7–5.5)
	Accurate passes (n)	418 ± 92	441 ± 79	$p = 0.669$	23 (−84–38)
	Instat (index)	195 ± 14.8	201.3 ± 13.4	$p < 0.001$	6 (−1.08–0.21)

Note: TD/min = total distance per minute; TSD/min = total sprint distance per minute; TD = Total distance; TSD = total sprint distance; ACC = acceleration; DECC = deceleration; min = minute; * significant differences between conditions ($p < 0.05$).

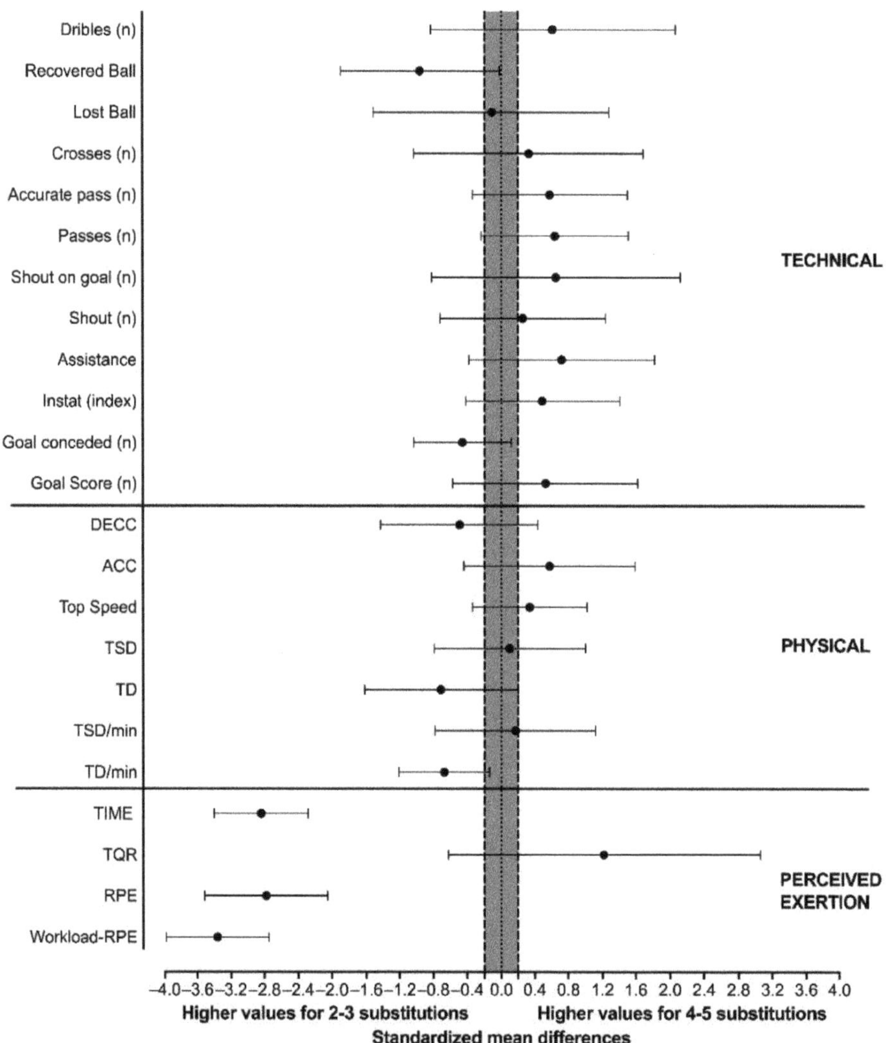

Figure 1. Standardized mean differences for the comparisons of the between substitutions conditions in all parameters of technical, physical, and perceived exertion; The grey area represents the smallest worthwhile difference which corresponds to a small effect size (0.2); Error bars represent the 95% confidence limits.

4. Discussion

This is the first study to investigate and compare the effect of the number of substitutions in-match (2–3 vs. 4–5) on physical and technical performance, and psychometrics measured (RPE, workload-RPE, and TQR) in elite women soccer players. The main findings were that 4–5 substitutions demonstrated a lower time played by player, lower RPE and workload-RPE, and higher TQR than 2–3 substitutions. In addition, no difference was observed in any physical or technical parameter of performance between conditions.

Interestingly, the present study demonstrated no differences between both substitution conditions in any physical performance (i.e., GPS metrics). Regardless of the number of substitutions used during the matches, the total metrics of the whole team did not change significantly. We analyzed the sum of metrics from all players, which induced a different perspective from previous studies that compared the relative physical demand (i.e., meters/minutes) among soccer players who played the entire match, were replaced, and substitutes [9,10]. For instance, Bradley et al. [9] demonstrated relative distance covered was inversely related to the time played, i.e., soccer players who covered 112 m/min during the entire match, replaced players who covered 116 m/min, and substitutes who traveled 120 m/min. These findings can be a consequence of fatigue or tactical decisions and adjustments (pacing strategy) in the energetic resources usage by soccer players who play longer periods to enable the completion of the matches [12]. For instance, the players can employ/adjust a pacing strategy that enables the entire match completion, that is, substitute players already know they will be exposed to a shorter match time, which allows them to cover higher distance/min than entire and replaced players [12]. From this perspective, it was conceivable that by increasing the number of substitutions the physical performance of the whole team would also be augmented. However, when we analyzed the sum of absolute values of all players and their mean relative values, using 4 or 5 substitutions did not elicit differences in match running performance compared to 2 and 3 substitutions. It is possible that the amount of distance covered by the team was only slightly affected and the difference did not reach statistical significance. Thus, our findings suggest that the new rule of substitutions may not have an impact on the physical performance of the team during the matches.

Nevertheless, we observed a significant difference in perceived internal load between conditions, according to RPE scale, the 2–3 substitutions were near the "maximal effort" (~8.1) and the 4–5 substitutions were near the "very hard" (~6.7) [18]. These facts can be explained by the fewer minutes played by soccer players during the matches in the 4–5 condition compared to 2–3 (65.5 vs. 76.4 min, respectively). Using five substitutions means that ~50% of the starting team was replaced (excluding the goalkeeper), which enables lower fatigue accumulated by players throughout the matches, mainly at the second half and the end of matches. Accordingly, the workload-RPE was lower in the 4–5 substitutions compared to 2–3 substitutions (441 vs. 623 A.U, respectively). Thus, the new rule of substitution did have a reduction in the perceived exertion of women soccer players. Interestingly, although no difference was observed in the physical performance, results indicate that internal load was reduced. Considering that RPE is affected not only by the intensity but also by the exercise volume [21], this finding seems a consequence of the lower RPE indicated by the substitute and replaced players, who played a shorter time, compared to those that played the entire match. It is reasonable to assume that soccer players who indicate lower RPE have improved wellness and consequently reduced the occurrence of injury and illness, as was reported by previous studies [15,22–24].

The shorter time played and lower workload-RPE experienced by substitutes and replaced players are possibly responsible for a superior recovery status after four and five substitutions compared to two and three substitutions. In this regard, higher TQR values suggested a faster recovery after matches. Moreover, Selmi et al. [25] observed correlation of 0.67 between TQR values (before each training session/match) and the successful passes in professional soccer players. Indeed, Kentta and Hassmen [17] indicated that recovery is crucial to avoid maladaptive physical and psychological effects of fatigue. For instance, Brink et al. [16] demonstrated that elite soccer players with higher workload-RPE presented higher injury rates and got sick more frequently. Furthermore, the occurrence of illness seems to be related to worst psychosocial stress and recovery after soccer training and matches. Moreover, a recent meta-analysis demonstrated that a soccer match alters the levels of muscle-injury markers, inflammation, and immunological cell tracking, impairs physical performance, and exacerbates perceptual responses until at least 72 h post-match [13]. Thus, using strategies to improve the balance between stress and recovery status (i.e., increasing the number of substitutions) may improve wellness and mitigate injuries and illnesses in soccer players. These outcomes may help the coaching staff to pre-plan new strategies for training and matches, mainly in the congested period, to improve the recovery for the next match, consequently decreasing the risk of injury and illness.

In line with previous studies [3,8], the technical parameters did not seem to be affected by the physical performance. However, the Instat index of the team was higher (no significant) in 4–5 substitutions compared to 2–3 substitutions (195 vs. 201.3, respectively). The Instat index in soccer is based on wide range of team- and individual-technical statistics. Modric et al. [19] suggested that a higher index demonstrates superior overall technical performance and that the Instat index may be related to the final game outcome (loss, draw, or win). It is conceivable that players with lower fatigue levels during the matches (i.e., substitute players) perform more efficient technical actions than those with higher fatigue (i.e., replaced players or those who played the entire match). For instance, Ferraz et al. [26] showed the negative influence of fatigue upon ball velocity in soccer kicking, which can be related with the impaired neuromuscular performance and may negatively affect coordination skills during the soccer matches. Although we did not observe improved technical performance with higher number of substitutions, more studies are needed to investigate these effects.

Finally, this study is not without limitations. First, it is a case study of only one soccer team. Although we are uncertain if the same findings can be generalized to all soccer teams in all leagues, it provides good evidence that increasing the number of substitutions enhances team recovery and decreases the perception of exertion. Second, this study was restricted to compare 2–3 and 4–5 substitutions because there were no data with no or one substitution during the matches. However, this result provides some evidence on the effect of the number of substitutions on physical and technical performance, perceived exertion, and recovery in women soccer players. Therefore, using four or five substitutions positively affected the balance between stress and recovery from elite women soccer players. Third, although there is a suggestion that injury rate can be influenced by workload, our results indicate lower overall workload in these elite players. Future studies should investigate the effect of the number of substitutions in the incidence of injuries during the matches [24,27]. In summary, the provisory rule of substitutions by FIFA brings a new insight for the future of soccer games.

5. Conclusions

The number of substitutions used during the matches did not influence the physical and technical performance of elite women soccer players. However, increasing the number of substitutions decreases the time played and the internal load (i.e., RPE and workload-RPE), which may have positive consequences on the recovery status (i.e., TQR) and the health of athletes by changing the balance between stress and recovery after matches. Therefore, the new provisory rule of substitutions provides some evidence to support the use of more soccer players during the match to improve the balance between stress and recovery.

Author Contributions: Conceptualization, R.K. and R.B.; methodology, L.C., R.S. and G.P.R.; Software, N.G., L.C. and V.C.; Formal analysis, R.S. and A.S.; Data Curation, A.S., V.C. and N.G.; investigation, R.K., writing—original preparation, R.K,, R.B, G.P.R. and R.A.; writing—review and editing, R.K., R.B. and R.A.; supervision, R.K. and R.B. All authors have read and agreed to the published version of the manuscript.

Funding: The authors are grateful to Sport Club Corinthians Paulista for allowing the development of this research and to CNPq for financial support (140606/2020-3).

Institutional Review Board Statement: All procedures conformed to the standards of the Declaration of Helsinki and were approved by the Ethics Committee of the University of Campinas (number: 4.497.044).

Informed Consent Statement: Informed consent was obtained from all subjects involved in the study.

Data Availability Statement: The data presented in this study are available on request from the corresponding author.

Conflicts of Interest: The authors declare no conflict of interest.

References

1. Bangsbo, J.; Mohr, M.; Krustrup, P. Physical and metabolic demands of training and match-play in the elite football player. *J. Sports Sci.* **2006**, *24*, 665–674. [CrossRef] [PubMed]
2. Krustrup, P.; Mohr, M.; Ellingsgaard, H.; Bangsbo, J. Physical demands during an elite female soccer game: Importance of training status. *Med. Sci. Sports Exerc.* **2005**, *37*, 1242–1248. [CrossRef] [PubMed]
3. Mohr, M.; Krustrup, P.; Andersson, H.; Kirkendal, D.; Bangsbo, J. Match activities of elite women soccer players at different performance levels. *J. Strength Cond. Res.* **2008**, *22*, 341–349. [CrossRef] [PubMed]
4. Ramos, G.P.; Nakamura, F.Y.; Penna, E.M.; Wilke, C.F.; Pereira, L.A.; Loturco, I.; Capelli, L.; Mahseredjian, F.; Silami-Garcia, E.; Coimbra, C.C. Activity Profiles in U17, U20, and Senior Women's Brazilian National Soccer Teams During International Competitions: Are There Meaningful Differences? *J. Strength Cond. Res.* **2019**, *33*, 3414–3422. [CrossRef]
5. Datson, N.; Drust, B.; Weston, M.; Jarman, I.H.; Lisboa, P.J.; Gregson, W. Match physical performance of elite female soccer players during international competition. *J. Strength Cond. Res.* **2017**, *31*, 2379–2387. [CrossRef]
6. Hewitt, A.; Norton, K.; Lyons, K. Movement profiles of elite women soccer players during international matches and the effect of opposition's team ranking. *J. Sports Sci.* **2014**, *32*, 1874–1880. [CrossRef]
7. Ramos, G.P.; Nakamura, F.Y.; Pereira, L.A.; Junior, W.B.; Mahseredjian, F.; Wilke, C.F.; Garcia, E.S.; Coimbra, C.C. Movement Patterns of a U-20 National Women's Soccer Team during Competitive Matches: Influence of Playing Position and Performance in the First Half. *Int. J. Sports Med.* **2017**, *38*, 747–754. [CrossRef]
8. Bradley, P.S.; Dellal, A.; Mohr, M.; Castellano, J.; Wilkie, A. Gender differences in match performance characteristics of soccer players competing in the UEFA Champions League. *Hum. Mov. Sci.* **2014**, *33*, 159–171. [CrossRef]
9. Bradley, P.S.; Lago-Penas, C.; Rey, E. Evaluation of the match performances of substitution players in elite soccer. *Int. J. Sports Physiol. Perform.* **2014**, *9*, 415–424. [CrossRef]
10. Padron-Cabo, A.; Rey, E.; Vidal, B.; Garcia-Nunez, J. Work-rate Analysis of Substitute Players in Professional Soccer: Analysis of Seasonal Variations. *J. Hum. Kinet.* **2018**, *65*, 165–174. [CrossRef]
11. Bradley, P.S.; Carling, C.; Archer, D.; Roberts, J.; Dodds, A.; Di Mascio, M.; Paul, D.; Diaz, A.G.; Peart, D.; Krustrup, P. The effect of playing formation on high-intensity running and technical profiles in English FA Premier League soccer matches. *J. Sports Sci.* **2011**, *29*, 821–830. [CrossRef] [PubMed]
12. Bradley, P.S.; Noakes, T.D. Match running performance fluctuations in elite soccer: Indicative of fatigue, pacing or situational influences? *J. Sports Sci.* **2013**, *31*, 1627–1638. [CrossRef] [PubMed]
13. Silva, J.R.; Rumpf, M.C.; Hertzog, M.; Castagna, C.; Farooq, A.; Girard, O.; Hader, K. Acute and Residual Soccer Match-Related Fatigue: A Systematic Review and Meta-analysis. *Sports Med.* **2018**, *48*, 539–583. [CrossRef]
14. Seshadri, D.R.; Thom, M.L.; Harlow, E.R.; Drummond, C.K.; Voos, J.E. Case Report: Return to Sport Following the COVID-19 Lockdown and Its Impact on Injury Rates in the German Soccer League. *Front. Sports Act. Living* **2021**, *3*, 604226. [CrossRef] [PubMed]
15. Drew, M.K.; Finch, C.F. The Relationship Between Training Load and Injury, Illness and Soreness: A Systematic and Literature Review. *Sports Med.* **2016**, *46*, 861–883. [CrossRef]
16. Brink, M.S.; Visscher, C.; Arends, S.; Zwerver, J.; Post, W.J.; Lemmink, K.A. Monitoring stress and recovery: New insights for the prevention of injuries and illnesses in elite youth soccer players. *Br. J. Sports Med.* **2010**, *44*, 809–815. [CrossRef]
17. Kentta, G.; Hassmen, P. Overtraining and recovery. A conceptual model. *Sports Med.* **1998**, *26*, 1–16. [CrossRef]
18. Foster, C.; Florhaug, J.A.; Franklin, J.; Gottschall, L.; Hrovatin, L.A.; Parker, S.; Doleshal, P.; Dodge, C. A new approach to monitoring exercise training. *J. Strength Cond. Res.* **2001**, *15*, 109–115.
19. Modric, T.; Versic, S.; Sekulic, D.; Liposek, S. Analysis of the association between running performance and game performance indicators in professional soccer players. *Int. J. Environ. Res. Public Health* **2019**, *16*, 4032. [CrossRef]
20. Hopkins, W.G.; Marshall, S.W.; Batterham, A.M.; Hanin, J. Progressive statistics for studies in sports medicine and exercise science. *Med. Sci. Sports Exerc.* **2009**, *41*, 3–13. [CrossRef]
21. Barroso, R.; Salgueiro, D.F.; do Carmo, E.C.; Nakamura, F.Y. The effects of training volume and repetition distance on session rating of perceived exertion and internal load in swimmers. *Int. J. Sports Physiol. Perform.* **2015**, *10*, 848–852. [CrossRef] [PubMed]
22. Gathercole, R.; Sporer, B.; Stellingwerff, T. Countermovement Jump Performance with Increased Training Loads in Elite Female Rugby Athletes. *Int. J. Sports Med.* **2015**, *36*, 722–728. [CrossRef]
23. Lopez-Valenciano, A.; Raya-Gonzalez, J.; Garcia-Gomez, J.A.; Aparicio-Sarmiento, A.; Sainz de Baranda, P.; De Ste Croix, M.; Ayala, F. Injury Profile in Women's Football: A Systematic Review and Meta-Analysis. *Sports Med.* **2021**, *51*, 423–442. [CrossRef] [PubMed]
24. Bengtsson, H.; Ekstrand, J.; Hagglund, M. Muscle injury rates in professional football increase with fixture congestion: An 11-year follow-up of the UEFA Champions League injury study. *Br. J. Sports Med.* **2013**, *47*, 743–747. [CrossRef] [PubMed]
25. Selmi, O.; Marzouki, H.; Ouergui, I.; BenKhalifa, W.; Bouassida, A. Influence of intense training cycle and psychometric status on technical and physiological aspects performed during the small-sided games in soccer players. *Res. Sports Med.* **2018**, *26*, 401–412. [CrossRef]

26. Ferraz, R.; van den Tillaar, R.; Marques, M.C. The effect of fatigue on kicking velocity in soccer players. *J. Hum. Kinet.* **2012**, *35*, 97–107. [CrossRef]

27. Dellal, A.; Lago-Penas, C.; Rey, E.; Chamari, K.; Orhant, E. The effects of a congested fixture period on physical performance, technical activity and injury rate during matches in a professional soccer team. *Br. J. Sports Med.* **2015**, *49*, 390–394. [CrossRef]

32

International Journal of
*Environmental Research
and Public Health*

Article

Weekly Training Load across a Standard Microcycle in a Sub-Elite Youth Football Academy: A Comparison between Starters and Non-Starters

José E. Teixeira [1,2,*], Luís Branquinho [1,3], Ricardo Ferraz [1,4], Miguel Leal [3], António J. Silva [1,5], Tiago M. Barbosa [1,2], António M. Monteiro [1,2] and Pedro Forte [1,2,3]

[1] Research Centre in Sports Sciences, Health and Human Development, 5001-801 Vila Real, Portugal
[2] Departamento de Desporto e Educação Física, Instituto Politécnico de Bragança, 5300-253 Bragança, Portugal
[3] Department of Sports, Higher Institute of Educational Sciences of the Douro, 4560-708 Penafiel, Portugal
[4] Department of Sports Sciences, University of Beira Interior, 6201-001 Covilhã, Portugal
[5] Department of Sports, Exercise and Health Sciences, University of Trás-os-Montes e Alto Douro, 5001-801 Vila Real, Portugal
* Correspondence: jose.eduardo@ipb.pt

Abstract: Compensatory training sessions have been highlighted as useful strategies to solve the differential weekly training load between the players' starting status. However, the influence of the players' starting status is still understudied in sub-elite youth football. Thus, the aim of this study was to compare the weekly training load on a standard microcycle in starters and non-starters of a sub-elite youth football academy. The weekly training load of 60 young sub-elite football players was monitored during a 6-week period using an 18 Hz global positioning system (GPS), 1 Hz telemetry heart rate, rating of perceived exertion (RPE), and total quality recovery (TQR). The total distance (TD) covered presented a significant difference between starters and non-starters with a moderate effect ($t = -2.38$, $\Delta = -428.03$ m, $p = 0.018$, $d = 0.26$). Training volume was higher in non-starters than in starter players ($TD_{Starters} = 5105.53 \pm 1684.22$ vs. $TD_{Non-starters} = 5533.56 \pm 1549.26$ m). Significant interactive effects were found between a player's starting status, playing time, and session duration in overall training load variables for within ($F = 140.46$; $\eta^2 = 0.85$; $p < 0.001$) and between-subjects ($F = 11.63$ to 160.70; $\eta^2 = 0.05$ to 0.76; $p < 0.001$). The player's starting status seems to only influence the training volume in sub-elite youth football, unless one considers the covariance of the playing time and session duration. Consequently, coaches should prioritize complementary training to equalize training volume and emphasize similar practice opportunities for non-starters. Future studies should evaluate the gap between training and match load, measuring the impact of recovery and compensatory sessions.

Keywords: workload; recovery; starting status; periodization; youth

Citation: Teixeira, J.E.; Branquinho, L.; Ferraz, R.; Leal, M.; Silva, A.J.; Barbosa, T.M.; Monteiro, A.M.; Forte, P. Weekly Training Load across a Standard Microcycle in a Sub-Elite Youth Football Academy: A Comparison between Starters and Non-Starters. *Int. J. Environ. Res. Public Health* **2022**, *19*, 11611. https://doi.org/10.3390/ijerph191811611

Academic Editors: Paul B. Tchounwou and Filipe Manuel Clemente

Received: 20 July 2022
Accepted: 13 September 2022
Published: 15 September 2022

Publisher's Note: MDPI stays neutral with regard to jurisdictional claims in published maps and institutional affiliations.

1. Introduction

Training load monitoring has been widely reported in youth football research [1,2]. Continuous training monitoring allows the measurement of the players' physical and physiological demands, allowing them to express their changes in performance and well-being [3,4]. Currently, analyzing and monitoring the weekly training load has become faster and easier to use due to advancements in tracking system applications [5,6]. Thus, the training representation and the game model can be quickly individually tailored through training load monitoring strategies [7,8]. Although most of the evidence has been produced in elite youth football, recently some studies have applied training load strategies in sub-elite cohorts [9–11]. Load variation over a standard microcycle in sub-elite football players seems to be influenced by week type, player's starting status, playing position, training mode, maturation status, and match-related contextual variables [1,11]. Previous studies have reported a high intra-week variation with a low inter-week variation across a standard microcycle in a sub-elite youth football team [10]. When comparing elite and sub-elite football contexts, several differences have been reported in training intensity and patterns [10,11]. However, the player's starting status is still poorly studied in sub-elite youth training, with few reports in elite contexts [12–15].

For these reasons, the need inevitably arises to make an adjustment to the training loads of starters and non-starters, as after a game some players may need a complete rest period [16] or regeneration [17], while others must follow their normal training schedule [16,17] or have complete compensatory sessions [18]. In this regard, a recent study indicates that it may be beneficial to use small-sided games (SSG) to control the imposed training load. In fact, even though players can perform the same type of SSG format, there seems to be evidence that the choice of training method (i.e., fractional or continuous) and recovery time between repetitions with the use of the fractional method results in increases and decreases in imposed training loads, respectively [17–19]. Based on the results of these authors, starters should perform continuous SSG formats to decrease training load responses, while non-starters should perform fractional formats with short recovery periods to increase training load responses, thus compensating for the difference in game load between players (compensatory training) during the weekly training microcycle [17]. In this way, SSG can be seen as a powerful tool to ensure that starter and non-starter players achieve the goals set by the coach for the training session (e.g., distances covered, different speed zones, accelerations, decelerations, heart rate among others) [1,17].

However, considering the above differences in competitive levels, it is important to determine the main contributing factors that influence the training load management [19,20]. From a long-term development perspective, managing physical qualities is an important factor in improving a player's future sporting career [21,22]. Load discrepancies based on starting status may require compensatory training sessions or competitive breaks optimization periods [23,24]. In professional football, Anderson et al. [13] described that the total activity volume (i.e., training and match load), as well as the total distance covered, were not different between starters, fringe players, and nonstarters, while Los Arcos et al. [14] stated that the match load was solely responsible for a higher weekly training load in starters compared to non-starters. Dalen and Lorås [12] reported a large amount of match-related high-speed running and sprint distances across the weekly training schedule for elite young football players. Therefore, the present research aims to examine the evidence-based training load and determine any similarities with the training of sub-elite youth football players. Thus, the main purpose of this study was to compare the weekly training load across a standard microcycle in starters and non-starters of a sub-elite youth football academy.

2. Materials and Methods

2.1. Participants and Study Design

Table 1 presents the baseline characteristics of the subsample of 60 male football players from a sub-elite Portuguese football academy. A total of 60 young football players aged between 13 and 20 years were analyzed in this prospective, observational, and cross-sectional study. The daily training load was continuously monitored during a 6-week period of the 2019–2020 competitive season. The training data corresponded to a total of 18 training sessions and 324 observation cases (i.e., starters and non-starters with 164 and 160 observations, respectively).

All participants were informed of the aims and risks of the research. The study only includes players whose legal guardian/next of kin had signed the informed consent to participate. The present research was conducted in accordance with the ethical standards of the Declaration of Helsinki. The experimental approach was approved and followed by the local Ethical Committee from the University of Trás-os-Montes e Alto Douro (3379-5002PA67807).

Table 1. Description of the participants' subsamples according to the player's starting status.

Variables	Starters (n = 164)	Non-Starters (n = 160)	Total (n = 324)
Age (y)	15.06 ± 1.85	15.33 ± 1.65	15.20 ± 1.75
Height (m)	1.74 ± 0.80	1.73 ± 0.69	1.73 ± 0.08
Weight (kg)	62.64 ± 10.46	62.31 ± 9.57	62.48 ± 10.03
BMI (Kg/m^2)	20.64 ± 2.01	20.58 ± 2.27	20.61 ± 2.14
Playing time (min)	73.82 ± 12.08	24.06 ± 9.67	49.25 ± 27.21
Session duration/wk (min)	148.13 ± 33.07	175.74 ± 43.52	161.77 ± 38.86

2.2. Eligibility Criteria for Training Data

The eligibility for training data was based on previous studies in sub-elite youth football [10,11] considering the following inclusion criteria: (a) young football players aged between 13 and 20 years old [1]; (b) at least five years of competitive experience in football [21]; (c) training files containing at least 35 consecutive minutes of playing time on the pitch [25]; (d) training data considered a competitive one-game per week schedule and complete full training sessions three times a week (~90 min) [10,11]. The exclusion criteria were: (a) total or partial absence from training due to data collection errors, injury events, rehabilitation sessions, individual training sessions, early withdrawal, and/or missing training; (b) football players aged under 13 or over 20 years; (c) the goalkeeper participated in the training session but was excluded from the analysis [1]. The exclusion criteria resulted in the elimination of 36 observation cases.

The players' starting status was divided into starters (i.e., started the game at least 55% of the games) and non-starters (i.e., started in less than 55% of the games) [13,26]. The average playing time was 73.82 ± 12.08 and 24.06 ± 9.67 min for starters and non-starters, respectively. The number of observations was adjusted by age group, specifically under 15 (U15), under 17 (U17), and under 19 (U19) [10,11]. The number of observations in weekly training data for each age was: U15 (n = 102), U17 (n = 99), and U19 (n = 120). The microcycle included three training sessions per week (~90 min) with the following "match day minus format" (MD): MD-3 (Tuesday), MD-2 (Wednesday), and MD-1 (Friday) [7,8]. The number of observations in weekly training data for each age was: MD-3 (n = 41), MD-2 (n = 38), and MD-1 (n = 44). The average training session consisted of 18 players with a training session and all age groups were trained on an outdoor pitch with official dimensions (FIFA standard; 100 × 70 m). The training sessions were performed on synthetic turf pitches, from 10:00 a.m. to 8:00 p.m., and with similar environmental conditions (14–20 °C; relative humidity 52–66%) [10,11].

2.3. Weekly Training Schedule

The sampled training sessions were categorized according to a specific focus, following the discussion with the coaching staff. All sampled training sessions started with a standard warm-up with low-intensity running, dynamic stretching for main locomotive lower limb muscles, technical actions, and ball possession. The overview of weekly training was potentially variable across categories, such as different training modes with an emphasis on game-based situations and sport-specific skills for football-specific exercises [27,28]. The typical weekly training schedule was categorized based on a typical training microcycle published on youth football [29,30].

The MD-3 (Tuesday) highlighted the recovery and technical skills with an emphasis on individual and group tactical actions by 1v1 to 6v6 small- and medium-sized games (SSG/MSG) (physiological set: 75–80% HR$_{max}$). The MD-2 (Wednesday) focused on the sectorial and collective tactical actions of the game model as training containing the use of large sided games (LSG) (i.e., 7v7 to 10v10) and simulated games (i.e., 11v11) with a physiological set of 75–80% HR$_{max}$. The MD-1 (Friday) emphasized goal-scoring situations and tactical schemes (i.e., corners, free-kicks, penalty kicks) (physiological set: 85–90% HR$_{max}$).

2.4. Procedures

The young sub-elite football players were monitored using a portable GPS throughout the whole training session duration (STATSports Apex®, Northern Ireland) [10,11]. The GPS device provides raw position velocity and distance at 18 Hz sampling frequencies, including an accelerometer (100 Hz), magnetometer (10 Hz), and gyroscope (100 Hz). Each player wore the micro-tech inner mini pocket of a custom-made vest supplied by the manufacturer, which was placed on the upper back between the two shoulder blades. All devices were activated 30 min prior to training data collection to allow clear and acceptable reception of the satellite signal. Respecting the optimal signal for the measurement of human movement, the match data considered eight available satellite signals as a minimum for the observations [31]. The validity and reliability of the global navigation satellite systems (GNSS) were guaranteed as the GPS has been well established in the literature [31–33]. The current variables and thresholds should consider a small error of around 1–2% reported in the 10 Hz STATSports Apex® units [31].

2.5. Training Load Measures

2.5.1. External Training Load

The external training loads were obtained with time–motion data: total distance (TD) covered (m), average speed (AvS), maximum speed (SPR) (m/s), relative high-speed running distance (rHSR) (m), high metabolic load distance (HMLD) (m), sprinting distance (SPR) (m), dynamic stress load (DSL) (a.u.), number of accelerations (ACC), and number of decelerations (DEC). The number and duration of sprints were also measured (SPR_D and SPR_N, respectively (m)). The GPS software provided information only on the locomotor categories above 5.50 m/s: rHSR (5.5–6.97 $km·h^{-1}$) and SPR (>6.97 $km·h^{-1}$). The sprints were measured by the number and average sprint distance (m). The HMLD is a metabolic variable defined as the distance in meters covered by a player when the metabolic power exceeds 25.5 $W·kg^{-1}$. HMLD variables include all high-speed running, accelerations, and decelerations above 3 m/s $m·s^{-2}$ [31–33]. Both acceleration variables (ACC/DEC) considered the number of accelerations and decelerations performed at maximum intensity (>3 and <3 m/s, respectively). The DSL variable was evaluated by a 100 Hz triaxial accelerometer integrated into the GPS device. The sum of the accelerations is presented in the three orthogonal axes of movement (X, Y, and Z planes) in arbitrary units (a.u.) [34]. The high-intensity activity thresholds were adapted from previous studies [1,2].

2.5.2. Internal Training Load

Heart Rate–Based Measures

Heart rate was recorded by a 1 Hz short-range telemetry system GARMIM TM HR band (International Inc., Olathe, KS, USA). Maximum heart rate (HR_{max}), average heart rate (AvHR), and percentage of HR_{max} (%HR_{max}) values were considered for analysis [35,36]. Training impulse was obtained by Akubat TRIMP [37], reporting a team TRIMP whose equation is based on individual data from the players' TRIMP; however, it was used to calculate the internal load for each player as: Akubat TRIMP = Training duration × $0.2053e^{3.5179x}$, among which the HR_{ratio} is the same in Banisters TRIMP [1], e = Napierian logarithms, 3.5179 is the e exponent, and x = HR_{ratio} [37]. HR_{max} was obtained by the Yo Yo intermittent recovery test level 1 (YYIR1) [38].

Perceived Exertion and Recovery

The perceived exertion was measured using the 15-point Portuguese Borg Rating of Perceived Exertion 6–20 Scale (Borg RPE 6–20) [39]. The sRPE was obtained by multiplying the total duration of training sessions for each individual RPE score (sRPE = RPE × session duration) following a scale from 6 to 20 [40]. To monitor recovery, each player was asked to report the total quality recovery (TQR) score on a scale from 6 to 20. This scale was proposed by Kenttä and Hassmén [41] to measure the athletes' recovery perceptions. RPE and TQR were individually collected approximately 30 min before and after each training session, respectively. Players were already familiarized with the procedures and the perceived data were collected using Microsoft Excel® spreadsheet (Microsoft Corporation, Redmond, WA, USA). Previous research has included both scales to examine perceived stress and fatigue in youth football [10,11].

2.6. Statistical Analysis

Robust estimates of a 95% confidence interval (CI) and data heteroscedasticity were calculated using randomly 1000 bootstrap samples [11,42]. Data are presented as the mean $\pm$ standard deviation (SD), mean differences (Δ) are presented in absolute values, and statistical significance was set at $p < 0.05$. Differences in the players' starting status were tested with an independent sample t-test [43]. Effect sizes (ES) were calculated based on Cohen's d and classified as: 0.2, trivial; 0.6, small; 1.2, large; and >2.0, very large [42,43]. A repeated-measure ANOVA was applied to compare the differences and interactive effects between playing time, session duration, and player's starting status in the weekly training load [44,45]. Data sphericity was checked by Mauchly's statistic, and where violated, a Greenhouse–Geiser adjustment was applied. For ANOVA, the ES was computed by the eta square (η^2) and interpreted as: $0 < \eta^2 \leq 0.04$, without effect; $0.04 < \eta^2 \leq 0.25$, minimum; $0.25 < \eta^2 \leq 0.64$, moderate; and $\eta^2 > 0.64$, strong [46,47]. A comparison of data visualization between starters and non-starters was performed by a violin diagram with a boxplot element (ggplot2). All statistical analyses and data visualization were conducted using JASP software (JASP Team, 2019; version 0.16.3, jasp-stats.org) [43].

3. Results

Weekly Training Load According to the Player's Starting Status

The descriptive statistics of weekly training load according to the player's starting status are presented in Table 2.

Table 2. Mean weekly training load according to the player's starting status.

Measures	Starters (*n* = 164)	Non-Starters (*n* = 160)
TD (m)	5105.53 $\pm$ 1684.22	5533.56 $\pm$ 1549.26
AvS (m/min)	48.07 $\pm$ 21.02	25.84 $\pm$ 16.00
SPR (m/s)	7.12 $\pm$ 1.38	7.52 $\pm$ 2.45
rHSR (m)	72.52 $\pm$ 77.88	81.53 $\pm$ 77.96
HMLD (m)	528.48 $\pm$ 289.14	588.58 $\pm$ 289.02
SPR_D (m)	41.26 $\pm$ 59.27	48.26 $\pm$ 57.39
SPR_N (n)	2.99 $\pm$ 3.51	3.49 $\pm$ 3.84
DSL (a.u.)	249.22 $\pm$ 130.66	252.30 $\pm$ 139.84
ACC (n)	44.14 $\pm$ 20.21	47.81 $\pm$ 22.83
DEC (n)	39.09 $\pm$ 20.97	43.85 $\pm$ 26.29
HR_{max} (bpm)	185.03 $\pm$ 10.00	186.89 $\pm$ 10.12
AvHR (bpm)	135.15 $\pm$ 11.04	136.78 $\pm$ 11.43
%HR_{max} (bpm)	72.87 $\pm$ 6.04	74.20 $\pm$ 6.13
Akubat TRIMP (a.u.)	86.05 $\pm$ 29.71	91.26 $\pm$ 34.07
RPE (a.u.)	12.99 $\pm$ 2.18	13.36 $\pm$ 2.18
sRPE (a.u.)	1169.45 $\pm$ 196.25	1202.06 $\pm$ 196.08
TQR (a.u.)	15.80 $\pm$ 2.17	15.99 $\pm$ 1.91

Abbreviations: ACC—acceleration; AvS—average speed; DEC—deceleration; HMLD—high metabolic load distance; RPE—ratings of perceived exertion; SPR—sprint distance; SPR_N—number of sprints; SPR_D—distance covered at sprinting; sRPE—session ratings of perceived exertion; TD—total distance; TQR—total quality recovery.

Table 3 presents the mean comparison between starters and non-starters for external and internal training loads. Only the TD covered presented a significant difference with a moderate effect when comparing between the player's starting status ($t = -2.38$, $\Delta = -428.03$ m, $p = 0.018$, $d = 0.26$). Training volume was higher for non-starters than starter players ($TD_{Starters} = 5105.53 \pm 1684.22$ vs. $TD_{Non\text{-}starters} = 5105.53 \pm 1684.22$ m). Neither the measures of external training intensity nor the internal training load showed significant differences. However, the high intensity showed a trend towards higher values in non-starters.

Table 3. Mean differences between starters and non-starters in the weekly training load.

Variables	*t*-Test				Cohen's *d*
Measures	**t**	**Δ**	**p**	**d**	**Qualitative Effect**
TD (m)	−2.38	−428.03	0.018	0.26	Moderate
AvS (m/min)	−1.88	−4.90	0.062	0.21	Moderate
SPR (m/s)	−1.81	−0.40	0.071	0.20	Moderate
rHSR (m)	−1.09	−9.43	0.277	0.12	Small
HMLD (m)	−1.87	−60.10	0.062	0.21	Moderate
SPR_D (m)	−1.08	−7.00	0.281	0.12	Small
SPR_N (n)	−1.23	−0.50	0.222	0.14	Small
DSL (a.u.)	−0.21	−3.08	0.838	0.02	Small
ACC (n)	−1.53	−3.67	0.126	0.17	Small
DEC (n)	−1.80	−4.76	0.072	0.20	Moderate
HR_{max} (bpm)	−1.67	−1.86	0.096	0.19	Small
AvHR (bpm)	−1.30	−1.63	0.193	0.15	Small
%HR_{max} (bpm)	−1.97	−1.33	0.049	0.22	Moderate
Akubat TRIMP (a.u.)	−1.47	−5.21	0.143	0.16	Small
RPE (a.u.)	−1.50	−0.36	0.136	0.17	Small
sRPE (a.u.)	−1.50	−32.61	0.136	0.17	Small
TQR (a.u.)	−0.86	−0.20	0.392	0.10	Small

Abbreviations: Δ—mean differences; ACC—accelerations; ALL—overall independent position group; AvS—average speed; bpm—beat per minute; CD—central defenders; CM—central midfielders; DEC—decelerations; FB—fullbacks; FW—forwards; rHSR—relative high speed running; SPR—sprints; TD—total distance; WM—wide midfielders.

When considering the playing time and session duration as co-variables, to compare the weekly training load in starters and non-starters, there were significant interactive effects between players' starting status, playing time, and session duration in overall training load variables, either for within-subjects (F = 140.46; $\eta^2 = 0.85$; $p < 0.001$) or for between-subjects (F = 11.63 to 160.70; $\eta^2 = 0.05$ to 0.76; $p < 0.001$). Figure 1 shows the comparison between starters and non-starters for each training load measure.

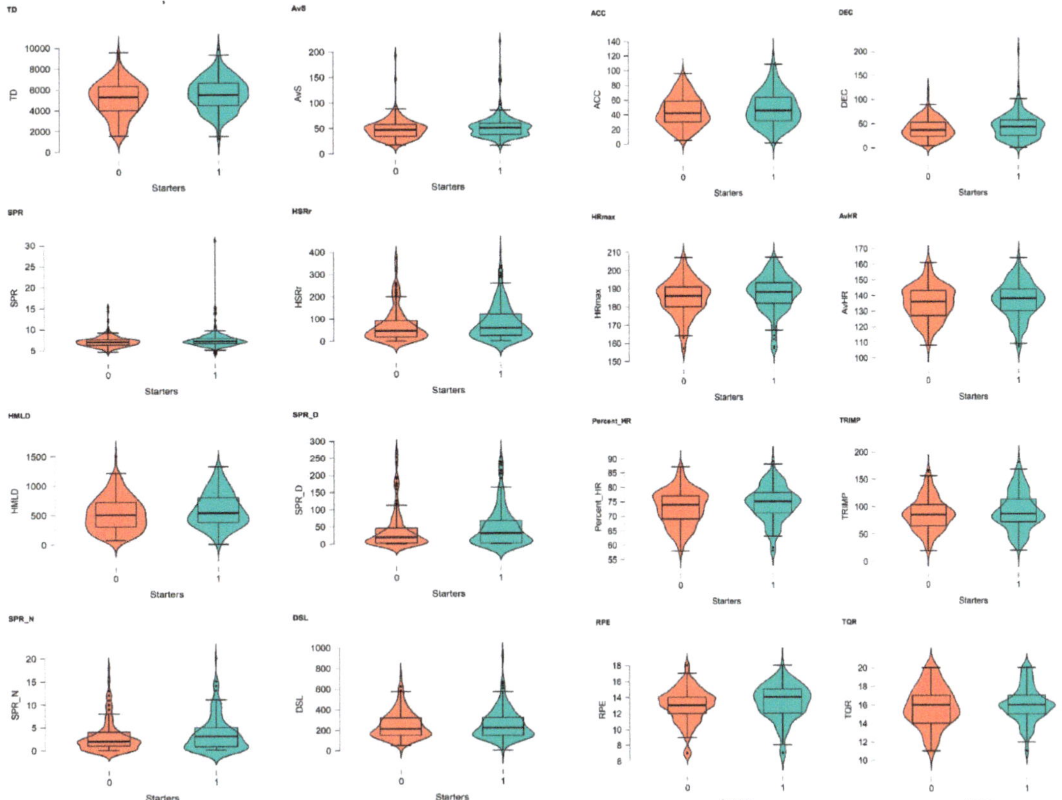

Figure 1. Comparison between starters and non-starters for each training load measure. Note: "Starters" coded 1 (red graph) and "non-starters" coded 2 (green graph).

4. Discussion

The main objective of this study was to compare the weekly training load across a standard microcycle in starters and non-starters of a sub-elite youth football academy. In general, the presented data suggested a trend towards a higher weekly training load in non-starting football players. Additionally, the external and internal training intensity did not seem to differ between the starting status of sub-elite youth football players. However, when considering the co-variance of the playing time and session duration, a significant interactive effect between the players' starting status, playing time, and session was reported in the overall training load variables.

In this study, only the TD covered seems to be influenced by the player's starting status in the young sub-elite, with a higher training volume for non-starters compared to starters (moderate effect). A possible explanation may be that coaches tend to prioritize complementary training to equalize training volume and emphasize similar practice opportunities for non-starters [23,24]. The fact that this sub-elite academy of training football only trains three times a week may represent that one of them might represent recovery training for the starters and compensatory training for the non-starters. The current findings are contrary to the evidence produced on the influence of the player's starting status for elite youth training. In youth elite football, Dalen and Lorås [12] determined a higher average weekly physical load for starters than non-starters in total covered distance, Banister's TRIMP, accelerations, and sprints. Furthermore, starters completed more moderate and high-intensity running than non-starters and fringe players in professional football [13].

Both training load analyses were performed during the in-season phase as in the present study [12,13]. On the contrary, this study determined that the non-starters covered more distance across the standard microcycle than starters. Current research also suggests a trend towards high-intensity activity as current training data showed a tendency towards higher values in non-starters, specifically for DEC, HSR, and SPR. The weekly training load disparities between elite and sub-elite football players are due to expertise level, periodization strategy, and training content [48,49], considering that it is possible that shorter training duration in sub-elite contexts may lead coaches to prioritize equity of practice opportunities for non-starters [48]. Otherwise, the intra- and inter-individual variation training load may influence the perceived exertion, pacing strategies, and high-intensity demands [11]. In addition, previous studies have demonstrated that non-starter players tend to have higher training workloads, which may result in overreaching, overtraining syndrome, and poor performance [44,45]. This evidence may also be due, in part, to the influence of maturational and motor development factors on the weekly training load [10,11]. Most importantly, the weekly training load across a standard microcycle should consider the co-variance of the playing time and session duration. This is because a non-starter may have 45 min, as well as a starter, since the players' starting statuses were based on the percentage of started matches and not on the playing time [13,26]. However, this evidence moves in the same direction as the weekly in-season training load verified in professional football players by Los Arcos et al. [14]. According to the study by Los Arcos, although a greater tendency towards a higher perceived exertion-based load for the starters was observed, only the match load was identified as a major factor contributing to a higher weekly training load. In the present study, the perceived exertion tended to be higher for starters than non-starters, for RPE, sRPE, and TQR. Previous studies have demonstrated that the perceived exertion does not seem to show differences either in age group or in maturity status [11]. Given this, the same assumptions seem to occur when considering the player's starting status as an influential factor in the accumulated training load [1]. All HR-based measures showed no statistical differences between starters and non-starters. However, similar to external training intensity, internal training intensity tends to be higher in non-starters. More specifically, non-starters have higher values for HR_{max} and $\%HR_{max}$. Teixeira et al. [11] described higher HR_{max} and Akubat TRIMP in U17, as well as $\%HR_{max}$, RPE, and sRPE in U15 sub-elite football players. The current weekly training load showed no differences for Akubat TRIMP between starters and non-starters. Although HR-based measures continue to be useful for training load monitoring, the limitations of measuring high-intensity movements are highly dependent on anaerobic components that have been widely described in the literature [1,2]. The standardization of the application of TRIMP methods to youth sub-elite football players should be considered to alleviate these problems [37]. Additionally, there is a need to reduce the dimensionality of the biomechanical and physiological datasets for a better understanding of the training load [11].

The current study presents some limitations that should be taken into consideration when interpreting and extending the results. First, the training load analysis included only one sub-elite football academy, so the applicability of the results must consider this specificity. Second, quantifying a weekly training load across a standard microcycle should also consider other influencing factors such as periodization structure and match-related contextual factors [10,11,50]. However, the current analysis did not include match data and, consequently, training and match load relationships [1]. The difference between recovery and compensatory sessions from other training days was also not analyzed [10]. Moreover, the training load was extracted from a complete training session, so that in the future the different training exercises should be subdivided to assess the task constraints and modality (i.e., fractional or continuous) such as SSG, high-intensity interval training (HIIT), and simulated game situations [1,51]. Pacing strategies and collective behavior should be considered in future research when analyzing the role of the starting status in match load [20,26,49]. In addition, future research should consider the relationship between compensatory training sessions with match load in youth sub-elite football, as this is an emerging research topic that has not yet been explored in sub-elite training contexts. Additionally, it is still necessary to compare how the behavior of sub-elite and elite football players differs in specific training drills and constrained tasks [1,10,11]. The lack of access to raw positional data made it challenging to perform the fragmented analysis of the entire training session [49]; therefore, future research should focus on physical, physiological, and technical–tactical analysis with an emphasis on comparing starters and non-starters [49,51]. Hence, more analyses are needed for this purpose with a broader follow-up, given the small sample and size of this prospective, cross-sectional, and observational study design. Research on the weekly training load with an integrative performance perspective should also be considered, as key technical and tactical indicators were not explored in this analysis [49].

5. Conclusions

The current research suggests a trend toward a higher weekly training load in non-starters, contrary to the published literature to date. The player's starting status only seems to influence the training volume in sub-elite youth football, unless the covariance of the playing time and session duration are considered. Thus, coaches seem to prioritize complementary training to equalize training volume and emphasize similar practice opportunities for non-starters. Future studies should evaluate the gap between training and match load in this comparison between starters and non-starters.

Author Contributions: Conceptualization, J.E.T. and P.F.; data curation, J.E.T., L.B. and P.F.; formal analysis, R.F., M.L. and P.F.; funding acquisition, L.B. and P.F.; investigation, J.E.T. and L.B.; methodology, T.M.B., A.M.M. and P.F.; resources, J.E.T. and P.F.; software, J.E.T., L.B. and M.L.; supervision, A.J.S. and P.F.; validation, T.M.B., A.M.M. and P.F.; writing—original draft, J.E.T.; writing—review and editing, L.B., R.F., M.L., T.M.B., A.J.S., A.M.M. and P.F. All authors have read and agreed to the published version of the manuscript.

Funding: This project was supported by the National Funds through the FCT—Portuguese Foundation for Science and Technology (project UIDB04045/2021).

Institutional Review Board Statement: The study was conducted in accordance with the guidelines of the Declaration of Helsinki and approved by the institutional Ethical Committee from the University of Trás-os-Montes e Alto Douro (Doc2-CE-UTAD-2021).

Informed Consent Statement: Informed consent was obtained from all subjects involved in the current investigation.

Data Availability Statement: Data are available under request to the contact author.

Acknowledgments: The authors acknowledge all of the coaches and playing staff for cooperation during all collection procedures.

Conflicts of Interest: The authors declare no conflict of interest.

References

1. Teixeira, J.E.; Forte, P.; Ferraz, R.; Leal, M.; Ribeiro, J.; Silva, A.J.; Barbosa, T.M.; Monteiro, A.M. Monitoring accumulated training and match load in football: A systematic review. *Int. J. Environ. Res. Public Health* **2021**, *18*, 3906. [CrossRef] [PubMed]
2. Miguel, M.; Oliveira, R.; Loureiro, N.; García-Rubio, J.; Ibáñez, S.J. Load measures in training/match monitoring in soccer: A systematic review. *Int. J. Environ. Res. Public Health* **2021**, *18*, 2721. [CrossRef] [PubMed]
3. Oliveira, R.; Ceylan, H.; Brito, J.; Martins, A.; Nalha, M.; Mendes, B.; Clemente, F. Within-and between-mesocycle variations of well-being measures in top elite male soccer players: A longitudinal study. *J. Men Health* **2022**, *18*, 94. [CrossRef]
4. Miguel, M.; Oliveira, R.; Brito, J.P.; Loureiro, N.; García-Rubio, J.; Ibáñez, S.J. External Match Load in Amateur Soccer: The Influence of Match Location and Championship Phase. *Healthcare* **2022**, *10*, 594. [CrossRef] [PubMed]
5. Akenhead, R.; Nassis, G.P. Training load and player monitoring in high-level football: Current practice and perceptions. *Int. J. Sports Physiol. Perform.* **2016**, *11*, 587–593. [CrossRef]
6. Rago, V.; Brito, J.; Figueiredo, P.; Costa, J.; Barreira, D.; Krustrup, P.; Rebelo, A. Methods to collect and interpret external training load using microtechnology incorporating GPS in professional football: A systematic review. *Res. Sports Med.* **2019**, *28*, 437–458. [CrossRef]

7. Oliveira, R.; Brito, J.; Martins, A.; Mendes, B.; Calvete, F.; Carriço, S.; Ferraz, R.; Marques, M.C. In-season training load quantification of one-, two- and three-game week schedules in a top European professional soccer team. *Physiol. Behav.* **2019**, *201*, 146–156. [CrossRef]

8. Martin-Garcia, A.S.; Diaz, A.G.M.; Bradley, P.S.; Morera, F.; Casamichana, D. Quantification of a Professional Football Team's External Load Using a Microcycle Structure. *J. Strength Cond. Res.* **2018**, *32*, 3511–3518. [CrossRef]

9. Trecroci, A.; Milanović, Z.; Frontini, M.; Iaia, F.M.; Alberti, G. Physical performance comparison between under 15 elite and sub-elite soccer players. *J. Hum. Kinet.* **2018**, *61*, 209–216. [CrossRef]

10. Teixeira, J.E.; Forte, P.; Ferraz, R.; Leal, M.; Ribeiro, J.; Silva, A.J.; Barbosa, T.M.; Monteiro, A.M. Quantifying sub-elite youth football weekly training load and recovery variation. *Appl. Sci.* **2021**, *11*, 4871. [CrossRef]

11. Teixeira, J.E.; Alves, A.R.; Ferraz, R.; Forte, P.; Leal, M.; Ribeiro, J.; Silva, A.J.; Barbosa, T.M.; Monteiro, A.M. Effects of chronological age, relative age, and maturation status on accumulated training load and perceived exertion in young sub-elite football players. *Front. Physiol.* **2022**, *13*, 832202. [CrossRef]

12. Dalen, T.; Lorås, H. Monitoring Training and Match Physical Load in Junior Soccer Players: Starters versus Substitutes. *Sports* **2019**, *7*, 70. [CrossRef]

13. Anderson, L.; Orme, P.; Di Michele, R.; Close, G.L.; Milsom, J.; Morgans, R.; Drust, B.; Morton, J.P. Quantification of Seasonal-Long Physical Load in Soccer Players With Different Starting Status From the English Premier League: Implications for Maintaining Squad Physical Fitness. *Int. J. Sports Physiol. Perform.* **2016**, *11*, 1038–1046. [CrossRef]

14. Arcos, A.L.; Mendez-Villanueva, A.; Martínez-Santos, R. In-season training periodization of professional soccer players. *Biol. Sport* **2017**, *34*, 149–155. [CrossRef] [PubMed]

15. Nobari, H.; Silva, R.; Manuel Clemente, F.; Oliveira, R.; Carlos-Vivas, J.; Pérez-Gómez, J. Variations of external workload across a soccer season for starters and non-starters. *Proc. Inst. Mech. Eng. Part P J. Sports Eng. Technol.* **2021**, 17543371211039296. [CrossRef]

16. Fransson, D.; Vigh-Larsen, J.F.; Fatouros, I.G.; Krustrup, P.; Mohr, M. Fatigue Responses in Various Muscle Groups in Well-Trained Competitive Male Players after a Simulated Soccer Game. *J. Hum. Kinet.* **2018**, *61*, 85–97. [CrossRef]

17. Branquinho, L.C.; Ferraz, R.; Marques, M.C. The Continuous and Fractionated Game Format on the Training Load in Small Sided Games in Soccer. *Open Sports Sci. J.* **2020**, *13*, 81–85. [CrossRef]

18. Schuth, G.; Szigeti, G.; Dobreff, G.; Revisnyei, P.; Pasic, A.; Toka, L.; Gabbett, T.; Pavlik, G. Factors Influencing Creatine Kinase Response in Youth National Team Soccer Players. *Sports Health* **2021**, *13*, 332–340. [CrossRef]

19. Sanchez-Sanchez, J.; Hernández, D.; Martin, V.; Sanchez, M.; Casamichana, D.; Rodriguez-Fernandez, A.; Ramirez-Campillo, R.; Nakamura, F.Y. Assessment of the external load of amateur soccer players during four consecutive training microcycles in relation to the external load during the official match. *Mot. Rev. Educ. Fís.* **2019**, *25*, e101938. [CrossRef]

20. Teixeira, J.; Branquinho, L.; Leal, M.; Marinho, D.; Ferraz, R.; Barbosa, T.; Monteiro, A.; Forte, P. Modeling the Major Influencing Factor on Match Running Performance during the In-Season Phase in a Portuguese Professional Football Team. *Sports* **2022**, *10*, 121. [CrossRef]

21. Ford, P.; De Ste Croix, M.; Lloyd, R.; Meyers, R.; Moosavi, M.; Oliver, J.; Till, K.; Williams, C. The long-term athlete development model: Physiological evidence and application. *J. Sports Sci.* **2011**, *29*, 389–402. [CrossRef] [PubMed]

22. Wrigley, R.D.; Drust, B.; Stratton, G.; Atkinson, G.; Gregson, W. Long-term soccer-specific training enhances the rate of physical development of academy soccer players independent of maturation status. *Int. J. Sports Med.* **2014**, *35*, 1090–1094. [CrossRef] [PubMed]

23. Marques, A.; Travassos, B.; Branquinho, L.; Ferraz, R. Periods of Competitive Break in Soccer: Implications on Individual and Collective Performance. *Open Sports Sci. J.* **2022**, *15*, e1875399X2112141. [CrossRef]

24. Calderón-Pellegrino, G.; Gallardo, L.; Garcia-Unanue, J.; Felipe, J.L.; Hernandez-Martin, A.; Paredes-Hernández, V.; Sánchez-Sánchez, J. Physical Demands during the Game and Compensatory Training Session (MD + 1) in Elite Football Players Using Global Positioning System Device. *Sensors* **2022**, *22*, 3872. [CrossRef] [PubMed]

25. de Dios-Álvarez, V.; Suárez-Iglesias, D.; Bouzas-Rico, S.; Alkain, P.; González-Conde, A.; Ayán-Pérez, C. Relationships between RPE-derived internal training load parameters and GPS-based external training load variables in elite young soccer players. *Res. Sports Med.* **2021**, *30*, 1–16. [CrossRef] [PubMed]

26. Barreira, J.; Nakamura, F.Y.; Ferreira, R.; Pereira, J.; Aquino, R.; Figueiredo, P. Season Match Loads of a Portuguese Under-23 Soccer Team: Differences between Different Starting Statuses throughout the Season and Specific Periods within the Season Using Global Positioning Systems. *Sensors* **2022**, *22*, 6379. [CrossRef]

27. Abade, E.A.; Gonçalves, B.V.; Leite, N.M.; Sampaio, J.E. Time-motion and physiological profile of football training sessions performed by under-15, under-17 and under-19 elite Portuguese players. *Int. J. Sports Physiol. Perform.* **2014**, *9*, 463–470. [CrossRef] [PubMed]

28. Coutinho, D.; Gonçalves, B.; Figueira, B.; Abade, E.; Marcelino, R.; Sampaio, J. Typical weekly workload of under 15, under 17, and under 19 elite Portuguese football players. *J. Sports Sci.* **2015**, *33*, 1229–1237. [CrossRef] [PubMed]

29. Branquinho, L.; Ferraz, R.; Marques, M.C. 5-a-Side Game as a Tool for the Coach in Soccer Training. *Strength Cond. J.* **2021**, *43*, 96–108. [CrossRef]

30. Zurutuza, U.; Castellano, J.; Echeazarra, I.; Guridi, I.; Casamichana, D. Selecting Training-Load Measures to Explain Variability in Football Training Games. *Front. Psychol.* **2019**, *10*, 2897. [CrossRef]

31. Beato, M.; Coratella, G.; Stiff, A.; Iacono, A. Dello The validity and between-unit variability of GNSS units (STATSports apex 10 and 18 Hz) for measuring distance and peak speed in team sports. *Front. Physiol.* **2018**, *9*, 1288. [CrossRef]

32. Nikolaidis, P.T.; Clemente, F.M.; van der Linden, C.M.I.; Rosemann, T.; Knechtle, B. Validity and reliability of 10-Hz global positioning system to assess in-line movement and change of direction. *Front. Physiol.* **2018**, *9*, 228. [CrossRef] [PubMed]

33. Whitehead, S.; Till, K.; Weaving, D.; Jones, B. The use of microtechnology to quantify the peak match demands of the football codes: A systematic review. *Sports Med.* **2018**, *48*, 2549–2575. [CrossRef] [PubMed]

34. Beato, M.; De Keijzer, K.L.; Carty, B.; Connor, M. Monitoring fatigue during intermittent exercise with accelerometer-derived metrics. *Front. Physiol.* **2019**, *10*, 780. [CrossRef]

35. Branquinho, L.; Ferraz, R.; Travassos, B.; Marques, M.C. Comparison between continuous and fractionated game format on internal and external load in small-sided games in soccer. *Int. J. Environ. Res. Public Health* **2020**, *17*, 405. [CrossRef] [PubMed]

36. Branquinho, L.; Ferraz, R.; Travassos, B.; Marinho, D.A.; Marques, M.C. Effects of different recovery times on internal and external load during small-sided games in soccer. *Sports Health* **2021**, *13*, 324–331. [CrossRef]

37. Akubat, I.; Barrett, S.; Abt, G. Integrating the internal and external training loads in soccer. *Int. J. Sports Physiol. Perform.* **2014**, *9*, 457–462. [CrossRef]

38. Bangsbo, J.; Iaia, F.M.; Krustrup, P. The Yo-Yo intermittent recovery test : A useful tool for evaluation of physical performance in intermittent sports. *Sports Med.* **2008**, *38*, 37–51. [CrossRef]

39. Cabral, L.L.; Nakamura, F.Y.; Stefanello, J.M.F.; Pessoa, L.C.V.; Smirmaul, B.P.C.; Pereira, G. Initial validity and reliability of the Portuguese Borg rating of perceived exertion 6–20 scale. *Meas. Phys. Educ. Exerc. Sci.* **2020**, *24*, 103–114. [CrossRef]

40. Brink, M.S.; Nederhof, E.; Visscher, C.; Schmikli, S.L.; Lemmink, K.A.P.M. Monitoring load, recovery, and performance in young elite soccer players. *J. Strength Cond. Res.* **2010**, *24*, 597. [CrossRef]

41. Kenttä, G.; Hassmén, P. Overtraining and recovery: A conceptual model. *Sports Med.* **1998**, *26*, 1–16. [CrossRef]

42. Maughan, P.C.; MacFarlane, N.G.; Swinton, P.A. Quantification of training and match-play load across a season in professional youth football players. *Int. J. Sports Sci. Coach.* **2021**, *16*, 1169–1177. [CrossRef]

43. JASP Team. JASP (Version 0.16.3) [Computer Software]. 2022. Available online: https://jasp-stats.org/download/ (accessed on 6 July 2022).

44. Martins, A.D.; Oliveira, R.; Brito, J.P.; Loureiro, N.; Querido, S.M.; Nobari, H. Intra-Season Variations in Workload Parameters in Europe's Elite Young Soccer Players: A Comparative Pilot Study between Starters and Non-Starters. *Healthcare* **2021**, *9*, 977. [CrossRef] [PubMed]

45. Oliveira, R.; Palucci Vieira, L.H.; Martins, A.; Brito, J.P.; Nalha, M.; Mendes, B.; Clemente, F.M. In-Season Internal and External Workload Variations between Starters and Non-Starters—A Case Study of a Top Elite European Soccer Team. *Medicina* **2021**, *57*, 645. [CrossRef] [PubMed]

46. Barbosa, T.M.; Ramos, R.; Silva, A.J.; Marinho, D.A. Assessment of passive drag in swimming by numerical simulation and analytical procedure. *J. Sports Sci.* **2018**, *36*, 492–498. [CrossRef]

47. Hopkins, W.G.; Marshall, S.W.; Batterham, A.M.; Hanin, J. Progressive statistics for studies in sports medicine and exercise science. *Med. Sci. Sports Exerc.* **2009**, *41*, 3–13. [CrossRef]

48. Ferraz, R.; Forte, P.; Branquinho, L.; Teixeira, J.; Neiva, H.; Marinho, D.; Marques, M. The Performance during the Exercise: Legitimizing the Psychophysiological Approach. In *Exercise Physiology*; IntechOpen: Rijeka, Croatia, 2022. [CrossRef]

49. Teixeira, J.; Forte, P.; Ferraz, R.; Branquinho, L.; Silva, A.; Barbosa, T.; Monteiro, A. Methodological Procedures for Non-Linear Analyses of Physiological and Behavioural Data in Football. In *Exercise Physiology*; IntechOpen: Rijeka, Croatia, 2022. [CrossRef]

50. Teixeira, J.E.; Leal, M.; Ferraz, R.; Ribeiro, J.; Cachada, J.M.; Barbosa, T.M.; Monteiro, A.M.; Forte, P. Effects of match location, quality of opposition and match outcome on match running performance in a Portuguese professional football team. *Entropy* **2021**, *23*, 973. [CrossRef]

51. Coutinho, D.; Gonçalves, B.; Santos, S.; Travassos, B.; Wong, D.P.; Sampaio, J. Effects of the pitch configuration design on players' physical performance and movement behaviour during soccer small-sided games. *Res. Sports Med.* **2019**, *27*, 298–313. [CrossRef]

International Journal of
Environmental Research and Public Health

The Immunological and Hormonal Responses to Competitive Match-Play in Elite Soccer Players

Ryland Morgans [1,*], Patrick Orme [2], Eduard Bezuglov [1], Rocco Di Michele [3] and Alexandre Moreira [4]

1   Department of Sports Medicine and Medical Rehabilitation, Sechenov State Medical University, 119991 Moscow, Russia
2   Sport Science and Medical Department, Bristol City FC, Bristol BS3 2EJ, UK
3   Department of Biomedical and Neuromotor Sciences, University of Bologna, 40126 Bologna, Italy
4   Department of Sport, School of Physical Education and Sport, University of Sao Paulo, Sao Paulo 05508-270, Brazil
*   Correspondence: rylandmorgans@me.com

Abstract: This study aimed to examine the salivary immunoglobulin A (s-IgA) and salivary cortisol (s-Cort) responses to competitive matches in elite male soccer players. Data were collected for 19 players (mean ± SD, age: 26 ± 4 years; weight: 80.5 ± 8.1 kg; height: 1.83 ± 0.07 m; body-fat 10.8 ± 0.7%) from a Russian Premier League team throughout a 6-week period during the 2021–2022 season. Physical match loads were measured through an optical tracking system. s-IgA and s-Cort were assessed one day before each match (MD − 1), 60-min before kick-off, 30-min post-match, and 48-h post-match (MD + 2). At 60-min before kick-off, s-IgA values were lower than at MD − 1 (90% CI difference 15.7–71.3 µg/mL). Additionally, compared to 60-min before kick-off, s-IgA was higher at 30-min post-match (90% CI difference 1.8–57.8 µg/mL) and at MD + 2 (90% CI difference 5.4–60.5 µg/mL). At 30-min post-match, s-Cort was higher than at 60-min before kick-off (90% CI difference 4.84–7.86 ng/mL), while on MD + 2 s-Cort was higher than at 60-min before kick-off (90% CI difference 0.76–3.72 ng/mL). Mixed model regressions revealed that longer playing time and total distance covered, and higher number of high-intensity accelerations, involved smaller s-IgA differences between 30-min post-match and 60-min before kick-off, and between 60-min before kick-off and MD + 2. Additionally, greater high-intensity and sprint distances, and a higher number of high-intensity and maximal accelerations, involved smaller s-Cort differences between 60-min before kick-off and MD + 2. In conclusion, the present results demonstrate that using salivary monitoring combined with match load may be a useful tool to monitor individual mucosal immunity and hormonal responses to match-play and the subsequent recovery periods in elite soccer players.

Keywords: salivary cortisol; salivary immunoglobulin A; physical match performance; recovery; soccer

Citation: Morgans, R.; Orme, P.; Bezuglov, E.; Di Michele, R.; Moreira, A. The Immunological and Hormonal Responses to Competitive Match-Play in Elite Soccer Players. *Int. J. Environ. Res. Public Health* **2022**, *19*, 11784. https://doi.org/10.3390/ijerph191811784

Academic Editor: Paul B. Tchounwou

Received: 31 August 2022
Accepted: 16 September 2022
Published: 18 September 2022

Publisher's Note: MDPI stays neutral with regard to jurisdictional claims in published maps and institutional affiliations.

1. Introduction

The physiological demands of soccer performance have been extensively researched over the past several decades [1]. It is widely accepted that undertaking ~90 min of a soccer match induces significant disruption to bodily homeostatic parameters. The impact that this has on various physiological processes in the hours and days following match-play has also been researched in detail [2,3].

Various methods have been employed within research settings in an effort to quantify the physiological impact following soccer match-play. These methods include assessment of neuromuscular function [4], blood sampling [5], subjective questionnaires [6] and saliva sampling [7]. While these methods have been used effectively to highlight relationships between the physiological status of soccer players and training and match demands, there is a need to fully understand the profile of the response to elite competitive soccer match-play. This further understanding may allow practitioners to individualize the schedule and program of players to ensure full recovery following match-play, reducing the likelihood of injuries, and optimal physical preparation for upcoming matches to maximize subsequent performance.

From the aforementioned methods, saliva sampling methods have been employed to quickly screen players for stress and illness on a regular basis throughout the season [7–10]. Saliva sampling is a relatively simple and non-invasive method that provides practitioners with a variety of markers that can be used to understand players physiological status pre- and post-match. Previous research has outlined the use of salivary markers such as salivary immunoglobulin A (s-IgA) [7], cortisol (s-Cort) [11], and testosterone [12] in soccer players following match-play.

As outlined above, the stressors of soccer match performance result in disruption to the physiological status of players. Mortatti et al. [8] reported a decrease in s-IgA concentration, a marker of mucosal immunity, in elite U19 soccer players when regularly monitored in a series of seven matches over 20 days, which may leave players more susceptible to illness, specifically through upper respiratory tract infections. Indeed, Springham et al. [10] also identified a cross-season suppression of s-IgA in professional soccer players, which was related to players perceived fatigue, sleep quality and muscle soreness suggesting the need to adopt s-IgA monitoring to aid in the prescription of training load and recovery. Therefore, methods that may be able to provide practitioners with an objective understanding of immune system function, in particular for mucosal immunity, in the period following match-play may be able to minimize the number of training days lost to illness over the course of a season [13].

Cortisol is a steroid hormone, detectable in saliva [14], that reflects catabolic balance [15]. Previous research has reported acute increases in s-Cort post-match in a variety of athletic populations including soccer [15,16], rugby [17], and Australian Rules football (AFL) [18], and differing training methods [19], which may persist for between 24- and 75-h [15,16,20]. Soccer studies that have examined longitudinal s-Cort responses have reported elevated values during periods of increased workload [21] and a reduction in Testosterone: Cortisol ratio toward the end of the competitive season [22]. However, previous longitudinal investigations are limited by infrequent or missing data points [21,22], while studies with short sampling periods have failed to examine the effect of elite competitive match-play or quantified the relationship between physical match performance and objective immunological (s-IgA) and hormonal (s-Cort) markers during the post-match 48-h recovery period. Thus, the ability to accurately analyze acute player responses is diminished.

Morgans et al. [7] presented data that reported fluctuations in s-IgA to be sensitive to changes in the physical demands placed on soccer players as a result of changes in fixture scheduling at different time points across the season. Values for s-IgA were decreased during periods of condensed fixture schedules (2–3 matches per week) but returned to 'normal' baseline measures during regular fixture schedules (one match per week). Similar findings were presented by Mortatti et al. [8] when assessing changes in s-IgA during a period of congested fixtures (seven matches in 20 days). However, these authors found no change in s-Cort concentration during the same period. These authors also suggest that further investigation is required to better understand the potential relationship between s-Cort and the physical demands of elite soccer match-play.

Therefore, this unique investigation aims to examine the s-IgA and s-Cort responses to match-play of elite European soccer players across six competitive fixtures compared with baseline and pre-match values, and to compare if and how these responses differ between starters and non-starters. Furthermore, the study aims to quantify the relationship between physical match performance and objective immunological (s-IgA) and hormonal (s-Cort) markers during the post-match 48-h recovery period. It was hypothesized that elite soccer match-play would induce changes in s-IgA and s-Cort when compared with baseline and that these changes would be greater for starters versus non-starters.

2. Materials and Methods

2.1. Experimental Approach to the Problem

This study examined 19 elite male soccer players from the same team over a 6-week period during the second phase of the season. The participants had been playing soccer for a minimum of 10 years. Thirteen of the players used in this investigation were members of their respected national teams. The sample was initially recruited based on squad selection across six league matches (home matches ($n = 4$), away fixtures ($n = 2$)) in the 2021–2022 season. The sample was further sub-divided into starting players ($n = 10$) and non-starting players ($n = 9$). Participant data were only included in the analyses as starting player when time spent on the field exceeded 45-min of the match. Players were considered for inclusion as starting player if they completed, based on the inclusion criterion of 45-min playing time, in three (50%) or more of the examined matches. During a regular week, samples were obtained one day before each match (MD − 1), 60-min before kick-off on match-day, 30-min post-match and 48-h post-match (two days (MD + 2)). All samples were collected prior to breakfast in the morning period (09.30–10.30 a.m.) 1-h pre-training except on match-day. In the six examined matches, kick- off time was 2.00 p.m. ($n = 3$), 4.30 p.m. ($n = 1$), and 7.00 p.m. ($n = 2$). Sample collection time on match-day varied due to the official start of the match but was consistently 60-min prior to kick-off. In addition to saliva assessment, all match performance data was collated for analysis. Except on match-day, all participants were in a fasted state and required to abstain from food and caffeine products for a minimum of 2-h prior to the collection of saliva, and all salivary samples were collected at the same time of day for all participants (09.30–10.30 a.m.) to minimize the residual effect of exercise and circadian variations.

2.2. Participants

A total of 19 male outfield players (mean ± SD, age 26 ± 4 years; weight 80.5 ± 8.1 kg; height 1.83 ± 0.07 m; body-fat 10.8 ± 0.7%) were involved in the study. Players were classified by position and grouped accordingly: Center Defender (CD) $n = 5$, Wide Defender (WD) $n = 3$, Center Midfield (CM) $n = 7$, Wide Forward (WF) $n = 2$, and Center Forward (CF) $n = 2$. All data evolved as a result of employment in which players were routinely monitored over the course of the competitive season. Nevertheless, approval for the study from the club was obtained [23] and the study was performed in accordance with the Helsinki Declaration principles. Ethical approval was granted by the local Ethics Committee of Sechenov University (N 22-21 dated 12 December 2021). To ensure confidentiality, all data were anonymized before analysis. Participants were fully familiarized with the experimental procedures within this study due to the regular testing protocols implemented as part of the clubs' performance monitoring strategy. During the study, players were instructed to maintain normal daily food and water intake, and no additional dietary interventions were undertaken.

2.3. Procedures

The study period included saliva sampling and all match performance across a 6-week phase of the 2021–2022 season. The training sessions performed during the investigation were representative of a typical training micro-cycle implemented within elite European soccer, involving a periodized training week encompassing low, moderate, and high intensity sessions leading to competitive match-play. No player reported a soft tissue injury, illness or infection during the data collection period.

2.4. Salivary Sampling

Given that soccer match-play induces a reduction in s-IgA concentration that return to basal levels within 18-h [24], we reasoned that collection of samples 48-h post-match would allow us to ascertain the effects of the acute suppression in s-IgA concentration from that associated with more chronic levels of stress. The diurnal rhythm of cortisol typically sees the highest concentrations in early morning with decreases as the day progresses [25].

Thus, players provided saliva samples pre-breakfast approximately 60-min before training on MD − 1, 60-min before kick-off on match-day, 30-min post-match and pre-breakfast approximately 60-min on MD + 2.

Saliva samples were collected and analyzed from this cohort of players using the Soma OFC II collection kits in combination with real-time Lateral Flow Device (LFD), respectively. This method has been previously validated for oral fluid collection in the immunoassay of immunoglobulins in sports persons [26,27] and correlates well with other methods (enzyme-linked immunosorbent assay) adopted in the determination of s-IgA [9] and s-Cort [11,20,24]. In accordance with the manufacturer's guidelines, after thoroughly rinsing their mouths with water, un-stimulated saliva samples were obtained. Players were required to place an Oral Fluid Collector (OFC II; Soma Bioscience, Oxfordshire, UK) consisting of a synthetic polymer-based swab attached to a polypropylene volume adequacy indicator stem in their mouth. Participants were instructed to swallow any saliva present within the oral cavity before placing the collection device on top of the tongue. Once the OFC kits collect 0.5 ml (± 20%) of oral fluid (collection time typically in the range of 20–50-s), the volume adequacy indicator turned blue and the player then placed the swab into the buffer bottle. The bottle was then mixed by gentle inversion for a period of 1–2-min, and the collected sample was ready to be analyzed through an IgA/Cortisol Dual LFD and photometric LFD reader (Soma Bioscience, Wallingford, UK). For the LFD, two-to-three drops of saliva/buffer mix were added to the sample window of the LFD cassette. The liquid in turn then ran the length of the test strip through creating a control and test line visible in the test window. Scanning of the LFD took place 15-min after the sample was added, being a competitive assay, the test line intensity was inversely proportional to the s-IgA and s-Cort concentration in the sample analyzed. This method has been previously validated [26–28] against ELISA (r^2 = 0.78) in 208 samples collected from a cohort of English Premier League soccer players [28].

2.5. Physical Load

League physical match performance data were collected using a two-camera optical tracking system (InStat, Moscow, Russia) that was installed to record and examine the technical and physical match performance during competitive league fixtures. The matches were filmed using two full HD, static cameras positioned on the centre line of the field, not less than 3-metres from the field and 7-metres in height. A consistent 25 Hz format was provided. Data were linearly interpolated to 50 Hz, smoothed using a 5-point moving average and then down-sampled to 10 Hz, which allowed analysis of all player actions with and without the ball [29]. The installation process, reliability, and validity of InStat have been previously reported [29]. Physical performance was analyzed using the InStat Analysis Software System and exported to the Microsoft Excel software for further analyses. InStat provided written permission to allow all match data to be used for research purposes. The physical match activity profile included: time on pitch (min); total distance covered (km); high intensity distance (km; total distance covered 5.5–7 m/s); sprint distance (km; total distance covered >7 m/s); number of high-intensity accelerations (peak speed 5.5–7 m/s); number of maximal accelerations (peak speed >7 m/s).

2.6. Statistical Analysis

All data are presented as the mean ± SD. When appropriate, 90% confidence intervals (CI) were also shown. Data were analyzed with the software R, version 4.2.0 (R Foundation for Statistical Computing, Vienna, Austria). Linear mixed models, with random intercepts for individual players' and match IDs, were used to assess the differences between the mean s-IgA and s-Cort values at the examined time points (MD − 1, 60-min before kick-off, 30-min post-match, and MD + 2) in starters compared to non-starters. The sample 60-min before kick-off was taken as the reference category to which values of MD − 1, 30-min post-match, and MD + 2 values were compared. Additionally, linear mixed-effect regressions with random intercept for players' and match IDs were performed to examine the effect of playing time (min) and variables related to the match physical effort (distances covered and number of accelerations), on s-IgA and s-Cort, respectively. The s-IgA and s-Cort differences between post- (30-min after) and pre-match (60-min before), and the s-IgA and s-Cort differences between 48-h post- and pre-match, were taken as outcome variables. Effect sizes were calculated from the coefficients of linear mixed models as Cohen's d through the lme.dscore function from the EMAtools package [30]. The absolute d value was interpreted as very small (<0.2), small (0.2–0.5), medium (0.5–0.8), large (>0.8). For all analyses, statistical significance was set at $p < 0.10$ due to the relatively small number of examined matches.

3. Results

The mean and SD s-IgA at the examined time points are shown in Figure 1. Sixty minutes before kick-off, the mean s-IgA value was significantly ($p = 0.0108$) lower than MD − 1, with an estimated difference of 43.5 µg/mL (90% CI: 15.7 to 71.3; d = 0.26, small). Additionally, compared to 60-min pre-match, there was a significantly higher value of s-IgA 30-min post-match ($p = 0.083$; estimated difference 29.8 µg/mL (90% CI: 1.8 to 57.8; d = 0.17, very small) and 48-h post-match ($p = 0.051$; estimated difference 33.0 µg/mL (90% CI: 5.4 to 60.5; d = 0.19, very small). No significant differences were observed between starters and non-starters at any time point, and there was no significant group x time interaction ($p > 0.10$).

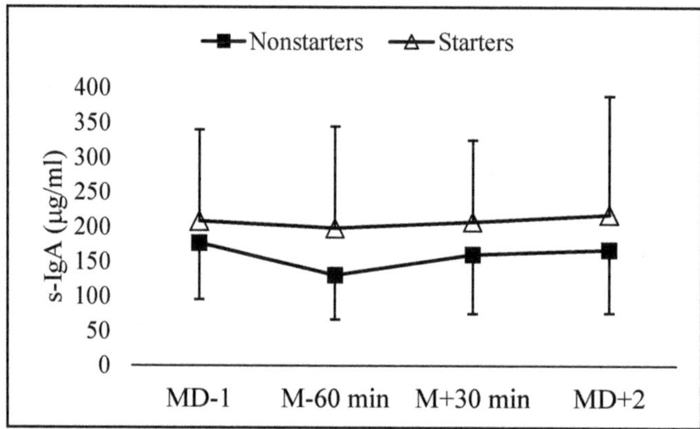

Figure 1. Mean and SD values of s-IgA the day before the match (MD − 1), 60-min before kick-off (M − 60 min), 30-min post-match (M + 30 min), and 48-h post-match (MD + 2).

Figure 2 shows the mean and SD values of s-Cort at the four examined time points. There was no significant difference between MD − 1 and 60-min before kick-off ($p = 0.118$). At 30-min post-match, s-Cort was significantly ($p < 0.001$) higher than 60-min pre-match, with an estimated difference of 6.35 ng/mL (90% CI: 4.84 to 7.86; d = 0.68, medium), while at 48-h post-match, s-Cort showed a decrease though it was still slightly higher ($p = 0.014$) than 60-min before kick-off, with an estimated difference of 2.47 ng/mL (90% CI: 0.76 to 3.72; d = 0.25 small). No differences were observed between starters and non-starters, and no significant time x group interaction was observed ($p > 0.10$).

Tables 1 and 2 shows the coefficients of fixed effects obtained with linear mixed model analysis with playing time and physical match performance variables as fixed factors, and individual values of s-IgA differences, 30-min post-match vs. 60-min before kick-off (Table 1), and 48-h post-match vs. 60-min before kick-off (Table 2), as outcome variables. These coefficients indicate the change in s-IgA differences post-match involved by a one-unit increase of the independent variable in that given match.

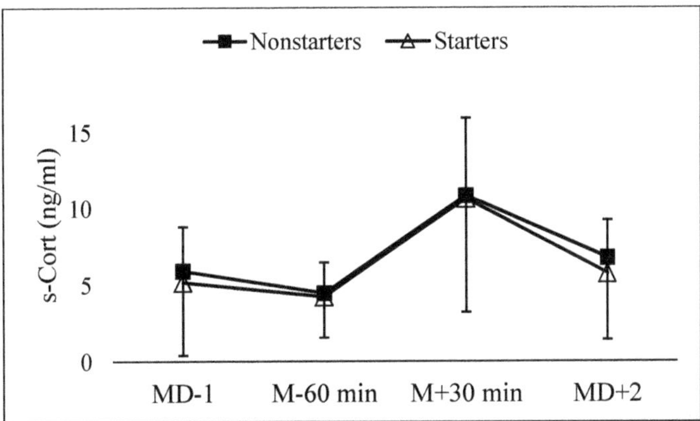

Figure 2. Mean and SD values of s-Cort the day before the match (MD − 1), 60-min before kick-off (M − 60 min), 30-min post-match (M + 30 min), and 48-h post-match (MD + 2).

Table 1. Effects of playing time and physical performance on s-IgA differences calculated between 30-min post-match and 60-min before kick-off time points.

	Coefficient (90% CI)	*p*-Value	Cohen's d
Playing time (min)	−0.74 (−1.36 to −0.12)	0.059 *	0.62
TD (km)	−6.82 (−12.36 to −1.31)	0.051 *	0.61
High-intensity distance (km)	−64.91 (−128.34 to −1.16)	0.102	0.43
Sprint distance (km)	61.99 (−200.82 to 320.77)	0.697	0.09
Number of high-intensity accelerations	−1.18 (−2.16 to 0.19)	0.057 *	0.48
Number of maximal accelerations	−0.42 (−4.89 to 3.98)	0.876	0.03

* *p* < 0.10. TD = Total distance; CI: Confidence Interval.

Table 2. Effects of playing time and physical performance on s-IgA differences calculated between 48-h post-match and 60-min before kick-off time points.

	Coefficient (90% CI)	*p*-Value	Cohen's d
Playing time (min)	−1.32 (−2.18 to −0.45)	0.013 *	0.80
TD (km)	−12.61 (−20.20 to −4.95)	0.007 *	0.81
High-intensity distance (km)	−125.65 (−211.21 to −37.23)	0.018 *	0.61
Sprint distance (km)	−323.09 (−679.52 to 49.44)	0.145	0.33
Number of high-intensity accelerations	−1.71 (−3.03 to −0.35)	0.037 *	0.52
Number of maximal accelerations	−3.16 (−9.20 to 3.18)	0.396	0.19

* *p* < 0.10. TD = Total distance; CI: Confidence Interval.

A 1-min longer time on pitch involved a 0.74 µg/mL smaller 30-min post-match/60-min before kick-off difference, with a medium effect (Table 1), and a 1.32 µg/mL smaller 48-h post-match/60-min before kick-off s-IgA difference, with a medium effect (Table 2). Similarly, a greater total distance covered and a higher number of high-intensity accelerations involved smaller s-IgA differences between 60-min before kick-off and 30-min or 48-h post-match, with d values ranging from medium to large (Tables 1 and 2). Additionally, greater high-intensity distance covered involved a smaller s-IgA difference between measurements taken 48-h post-match and 60-min before kick-off, with a medium effect (Table 2).

The fixed effects obtained from linear mixed models, with time on pitch and physical match performance variables as fixed factors, and individual values of s-Cort differences as outcome variables are presented in Table 3 (30-min post-match vs. 60-min before kick-off difference) and Table 4 (48-h post-match vs. 60-min before kick-off difference).

Table 3. Effects of playing time and physical performance on s-Cort differences calculated between 30-min post-match and 60-min before kick-off time points.

	Coefficient (90% CI)	*p*-Value	Cohen's d
Playing time (min)	0.005 (−0.042 to 0.053)	0.838	0.06
TD (km)	0.01 (−0.41 to 0.42)	0.999	0.00
High-intensity distance (km)	0.31 (−4.24 to 4.83)	0.910	0.03
Sprint distance (km)	1.09 (−17.42 to 18.78)	0.921	0.02
Number of high-intensity accelerations	−0.011 (−0.081 to 0.059)	0.778	0.07
Number of maximal accelerations	0.073 (−0.243 to 0.371)	0.686	0.09

TD = Total distance; CI: Confidence Interval.

Table 4. Effects of playing time and physical performance on s-Cort differences calculated between 48-h post-match and 60-min before kick-off time points.

	Coefficient (90% CI)	*p*-Value	Cohen's d
Playing time (min)	−0.029 (−0.065 to 0.007)	0.193	0.35
TD (km)	−0.30 (−0.62 to 0.01)	0.116	0.41
High-intensity distance (km)	−3.54 (−7.02 to −0.11)	0.097 *	0.38
Sprint distance (km)	−15.55 (−29.06 to −2.06)	0.063 *	0.43
Number of high-intensity accelerations	−0.0056 (−0.1098 to −0.0035)	0.086 *	0.40
Number of maximal accelerations	−0.236 (−0.462 to −0.009)	0.091 *	0.39

* $p < 0.10$. TD = Total distance; CI: Confidence Interval.

There was no significant effect of playing time, distances covered or the number of high-intensity or maximal accelerations on s-Cort differences between 30-min post-match and 60-min before kick-off (all $p > 0.10$) (Table 3). Conversely, greater high-intensity and sprint distances, and a higher number of high-intensity and maximal accelerations, involved smaller s-Cort differences between 48-h post-match and 60-min before kick-off, with small effects (Table 4).

4. Discussion

This investigation aimed to examine the s-IgA and s-Cort responses to match-play of elite soccer players across six competitive fixtures in the 2021–2022 season compared with baseline and pre-match values. Furthermore, the study aimed to quantify the relationship between physical match performance and objective immunological (s-IgA) and hormonal (s-Cort) markers during the post-match 48-h recovery period. One of the main findings of the present study was the significant though slight decrease in s-IgA concentration from MD − 1 to 60-min before kick-off. It is reasonable to suggest that this result is somewhat unexpected as the release of s-IgA is under strong neuroendocrine control [31], and the activation of the sympathetic nervous system associated with player's match preparation would, on the contrary, increase s-IgA concentration. Previously, it has been suggested that these mechanisms are responsible for the increases in s-IgA concentration induced by acute stress [32]. This result however, is unique in elite professional male soccer players and may suggest that psychological factors related to official match-play preparation may affect s-IgA concentration, and consequently, mucosal immune function. Moreira et al. [33], demonstrated in elite male volleyball players a significantly lower pre-match s-IgA concentration for a final championship match compared with pre-match s-IgA values for a regular season match. This result suggests that players' perceived importance of the match affect s-IgA concentration, highlighting therefore, the role of psychological factors in modulating the mucosal immunity in team-sport athletes. Indeed, this result further indicates that monitoring resting s-IgA in team-sports athletes would provide valuable information regarding how athletes cope with competition induced stress.

Regarding coping with stress related to competitive match preparation, the present results reported lower s-IgA concentration 60-min before kick-off compared to MD − 1, which may be partly explained by the well-known differences in responses to acute stress between active and passive coping strategies [34]. Bosch et al. [34] examined the acute immunological effects of two different laboratory stressors ("active coping" via a time-paced memory test and "passive coping" via a stressful video showing surgical operations). The results of the study showed that active coping led to increases in s-IgA concentration, while, passive coping induced a decrease in s-IgA concentration. Considering that the preparation for an official match may impose a significant psychological stress on team-sports athletes [33,35], it may therefore be inferred that the adoption of passive coping strategies before official match-play may negatively impact the mucosal immune function which in turn may increase the likelihood of upper respiratory tract infection occurrences. The present results in conjunction with the aforementioned data may possibly provide an opportunity for sport scientists and professionals working with soccer players to adopt active coping strategies during the preparation period for official soccer matches, and highlight the potential for the introduction of affective or positive emotional engagement. Further studies should focus on examining whether structured active coping tasks minimize the negative effect (i.e., decreasing s-IgA concentration) of the inherent stress associated with preparation for official match-play.

The current results also demonstrated an increasing trend in s-IgA concentration at 30-min and 48-h post-match, compared to 60-min before kick-off. These results suggest a short-term (acute) stress response induced enhancement of mucosal immune function [36]. Psychological and physiological stressors have been shown to stimulate biological stress. These responses are signals to cells and tissues, which express themselves as receptors for the released biological factors, leading therefore to the activation of all bodily systems, including the immune system. In contrast to chronic stress, that may lead to suppression or dysregulation of immune function, while impacting negatively the mucosal immunity [37], the present results suggest that the short-term stressors related to official soccer match-play may induce enhancement of immune function in professional soccer players. This is a positive response which prepares athletes for the imposed challenges associated with competition. It is important to highlight that previous studies have shown that factors such as corticosterone and epinephrine, released due to the presence of a stressor, are mediators of a short-term stress induced immuno-enhancement, while a variety of studies have shown increases or no changes in s-IgA concentration from pre- to post-match in team-sport athletes [9,33,38], professional female soccer players [39], and professional male soccer players [40]. Previous studies in soccer players demonstrated that elevated levels of psycho-physiological stress may negatively affect the mucosal immune function, with decreases in s-IgA concentration across periods of congested fixtures or intensive training loads [7,8,12,41]. Considering our results in combination with the existing literature, it could be reasonable to suggest that the probability of observing no changes or even increases in s-IgA concentration is high for acute stress (i.e., from pre- to post-match), while on the other hand, the chronic effect of accumulated stress, notably, when performing successive matches in a short period of time, may negatively affect the mucosal immunity of players.

The design of the present study allowed the observation of s-IgA responses to actual physical match load that have not yet been demonstrated in official soccer matches with elite professional male players. Despite the observed trend to increase s-IgA concentration from pre-match to 30-min and 48-h post-match, it is notable that, when performing a higher workload, players seemed to present a slower return to their initial s-IgA concentration. The 1-min longer playing time on pitch produced a 0.74 µg/mL smaller 30-min post-match/60-min before kick-off difference and a 1.32 µg/mL smaller 48-h post-match/60-min before kick-off s-IgA difference. Smaller s-IgA concentration differences between 60-min before kick-off and 30-min or 48-h post-match were also observed in association with greater total distance covered, and with a higher number of high-intensity accelerations. Additionally, greater high-intensity distance covered involved a smaller s-IgA difference between 48-h post-match and 60-min before kick-off. This unique and important finding of the present study suggests that an inverted-U/bell-shaped relationship may be observed between match-workload and the effects on mucosal immune function.

Thus, when performing higher workload, above a given threshold, players may be more prone to trivial increases or even reductions in s-IgA concentrations. In addition, this result may aid in explaining the increased likelihood of a suppressed effect from accumulated and successive match-play in s-IgA concentration, as this workload accumulation would affect plasma cells functions (immunoglobulin-secreting plasma cells) and the rate of IgA transcytosis across the epithelial cell. This result suggests a novel role for physical match workload monitoring and its impact on mucosal immunity in professional soccer players.

In relation to s-Cort, there was no significant difference between MD − 1 and 60-min before kick-off. This result suggests that the expected anticipatory stress response to match participation [42] did not occur. This finding might be associated with the high-level of the examined players and with their habitual lead-in process to cope with the pressure and anxiety involved in the period preceding the start of official matches. In this sense, van Paridon et al. [42] reported in their systematic review that the anticipatory stress response and cortisol reactivity, in both male and female athletes competing at international level, do not present a significant anticipatory cortisol response. Moreover, in earlier research, Alix-Sy et al. [43] despite showing a significant increase in s-Cort concentration at pre-match compared to a non-training day in professional French soccer players, reported a significant positive association between unpleasant somatic emotions and cortisol. Indeed, Alix-Sy et al. [43] also demonstrated no differences in s-Cort between starters and non-starters, as observed in the current study. Furthermore, it should be highlighted that in their study, the authors compared a non-training day with official matches, while in the present study, saliva collection occurred during habitual training sessions performed one day before matches. This difference may influence, at least in part, the present result of no change in s-Cort.

Considering these findings, we might suggest that the players evaluated in the present study did not show a s-Cort rise from MD − 1 to 60-min before kick-off possibly due to their positive evaluation of the potential match challenges, which in turn may be related to their perception of relative situational control and the non-decisive nature of regular season matches. Considering the results of the present study it may be suggested that due to the nature of the evaluated matches and the level of the assessed players, the s-Cort anticipatory responses (MD − 1 vs. 60-min before kick-off) indicated an optimal cognitive and behavioural player state to participate in the matches.

As expected, a significant increase in s-Cort from 60-min before kick-off to 30-min post-match was observed, while at 48-h post-match, s-Cort showed a decrease though still slightly higher than 60-min before kick-off, and no differences were observed between starters and non-starters. The increases in s-Cort reinforces that official soccer match-play induce significant psychophysiological stress likely related to physical demands associated with the volume and intensity of match-play, leading to increased secretion of s-Cort, as also reported in A-League [16] and intercollegiate soccer players [44]. It is important to highlight that the psychological factors involved in official match-play may play a role in this result. These results in conjunction allow us to infer that besides the well-known effect of increased s-Cort related to exercise stress, which represents per se a potent physiological stressor [45], the pressure of official match-play may be considered as an additional stress factor, possibly due to its social-evaluative task characteristics combined with other contextual factors inherent to sports competition as proposed by Arruda et al. [46].

Indeed, as demonstrated more recently by Rowell et al. [16], a substantial individual variability in s-Cort response to soccer match-play may be expected, including the responses within 48-h post-match. Furthermore, the psycho-physiological relationships and the impact of situational factors have been reported to influence cortisol responses to match-play in soccer players [47]. Thus, the present results add to the literature and suggest that contextual factors other than being a starter or non-starter may influence the variability in players s-Cort responses. The uniqueness of the present study allowed us to examine the effect of match-load measures on s-Cort time-course responses. A novel and interesting finding of the present study was that the greater high-intensity distance, sprint distance, number of high-intensity and maximal accelerations performed, the smaller the s-Cort differences between 48-h post-match and 60-min before kick-off. This result suggests that performing a greater number of high-intensity actions during the match would increase the associated stress, that in turn may hinder s-Cort recovery to resting values. Moreover, the present results suggest that high-intensity distance, sprint distance, number of high-intensity and maximal accelerations may be employed as reliable markers of individual external match-load inducing stress, and possibly predict catabolic state induced by match-play, rather than dividing players into starters and non-starters groups. In addition, the findings highlight the need to monitor in conjunction with the individual external match-load and the s-Cort response of players to account for individual variability in recovery from match-play. The results also suggest that examining s-Cort responses from pre-match to 30-min post-match might not aid in observing true changes in s-Cort during the recovery time-course.

Despite the interesting findings of the current study, some limitations should be acknowledged. Firstly, our study only focused on one elite European professional soccer team across a 6-week period, and as a result, the findings and practical implications must be considered with caution when applying to another set of players from a league with different characteristics such as match demands, travel [48], climate [49], and over an extended period of time during a different phase of the season (early-, mid-, late-season, congested Christmas schedule). Furthermore, the sample size was also a limitation due to this study being conducted in the real-world, conducted with players from an elite soccer club. Our sample was selected as a convenience sample by recruiting all available outfield players from the first team of the club involved. Nevertheless, similar sample sizes have been used in previous studies conducted in elite soccer players in this research field. Secondly, match outcome was not considered, which has the potential to affect immunological and hormonal recovery profiles. Future investigations are warranted to evaluate these factors as they may be particularly relevant in different leagues across varying athletic populations during the season. Other limitations include the absence of training load, fatigue, and fitness profiling data [50].

Practical Implications

The present findings may provide practitioners with detailed knowledge about acute and chronic variations in physical match performance and the subsequent recovery responses, that can be practically useful to assess and interpret change in individual and team performance. Previously, a number of practical recommendations to monitor immune function in athletes have been documented [7,9,10,16,51]. Match-play with higher physical outputs did not necessarily produce disturbances to mucosal immunity and hormonal balance. Therefore, accordingly, designing a structured, planned and individualized tailored recovery strategy and potential for squad rotation should be considered during demanding stages of the season to ensure immunological and hormonal recovery. Previous results highlighted that this might be particularly important during congested fixture schedules (Christmas fixture period) [7] and toward the end of the season [10]. Our findings support the use of s-IgA and s-Cort monitoring in professional soccer players and devising individual thresholds to determine values associated with inadequate recovery.

5. Conclusions

As a result of this specific investigation, the data demonstrate for the first time that the use of salivary monitoring in combination with physical match load may be a useful tool to monitor individual mucosal immunity and hormonal responses to elite soccer match-play and the subsequent recovery periods. However, surprisingly no significant differences were observed between starters and non-starters at any time point, thus additional research is required. Finally, analysis of specific time points during recovery also warrants further investigation.

Author Contributions: Conceptualization, R.M. and R.D.M.; methodology, R.M.; software, R.D.M.; validation, R.M. and R.D.M.; formal analysis, R.D.M.; investigation, R.M. and E.B.; resources, R.M. and E.B.; data curation, R.M.; writing—original draft preparation, R.M., P.O., R.D.M. and A.M.; writing—review and editing, R.M., E.B., R.D.M. and A.M.; visualization, R.D.M.; supervision, R.M.; project administration, R.M. All authors have read and agreed to the published version of the manuscript.

Funding: This research received no external funding.

Institutional Review Board Statement: The study was conducted according to the guidelines of the Declaration of Helsinki, and approved by the local Ethics Committee of Sechenov University (N 22-21 dated 12 December 2021).

Informed Consent Statement: Informed consent was obtained from all subjects involved in the study.

Data Availability Statement: The data are not publicly available due to privacy reasons.

Acknowledgments: The authors want to thank all the players, and medical staff involved in the study for the professionalism shown throughout. The authors also want to state that the results of the present study do not constitute endorsement of any products.

Conflicts of Interest: The authors declare no conflict of interest.

References

1. Bangsbo, J. Physiological demands of football. *Sports Sci. Exch.* **2014**, *27*, 1–6.
2. Nedelec, M.; McCall, A.; Carling, C.; Legall, F.; Berthoin, S.; Dupont, G. Recovery in soccer, Part 1—Post match fatigue and time course of recovery. *Sports Med.* **2012**, *42*, 997–1015. [PubMed]
3. Nedelec, M.; McCall, A.; Carling, C.; Legall, F.; Berthoin, S.; Dupont, G. Recovery in soccer, Part 2—Recovery strategies. *Sports Med.* **2013**, *43*, 9–22. [CrossRef] [PubMed]
4. Silva, J.R.; Ascensão, A.; Marques, F.; Seabra, A.; Rebelo, A.; Magalhães, J. Neuromuscular function, hormonal and redox status and muscle damage of professional soccer players after a high-level competitive match. *Eur. J. Appl. Physiol.* **2013**, *113*, 2193–2201. [CrossRef] [PubMed]
5. Silva, J.R.; Rebelo, A.; Marques, F.; Pereira, L.; Seabra, A.; Ascensão, A.; Magalhães, J. Biochemical impact of soccer: An analysis of hormonal, muscle damage, and redox markers during the season. *Appl. Physiol. Nutr. Metab.* **2014**, *39*, 432–438. [CrossRef]
6. Brink, M.S.; Visscher, C.; Arends, S.; Zwerver, J.; Post, W.J.; Lemmink, K.A. Monitoring stress and recovery: New insights for the prevention of injuries and illnesses in elite youth soccer players. *Br. J. Sports Med.* **2010**, *44*, 809–815. [CrossRef]
7. Morgans, R.; Orme, P.; Anderson, L.; Drust, B.; Morton, J.P. An Intensive Winter Fixture Schedule Induces a Transient Fall in Salivary IgA in English Premier League Soccer Players. *Res. Sports Med.* **2014**, *22*, 346–354. [CrossRef]
8. Mortatti, A.L.; Moreira, A.; Aoki, M.S.; Crewther, B.T.; Castagna, C.; de Arruda, A.F.S.; Filho, J.M. Effect of Competition on Salivary Cortisol, Immunoglobulin A, and Upper Respiratory Tract Infections in Elite Young Soccer Players. *J. Strength Cond. Res.* **2012**, *26*, 1396–1401. [CrossRef]
9. Morgans, R.; Owen, A.; Doran, D.; Drust, B.; Morton, J.P. Prematch Salivary Secretory Immunoglobulin A in Soccer Players from the 2014 World Cup Qualifying Campaign. *Int. J. Sports Physiol. Perform.* **2015**, *10*, 401–403. [CrossRef]
10. Springham, M.; Williams, S.; Waldron, M.; Strudwick, A.J.; Mclellan, C.; Newton, R.U. Salivary Immunoendocrine and Self-report Monitoring Profiles across an Elite-Level Professional Football Season. *Med. Sci. Sports Exerc.* **2020**, *53*, 918–927. [CrossRef]
11. Fothergill, M.; Wolfson, S.; Neave, N. Testosterone and cortisol responses in male soccer players: The effect of home and away venues. *Physiol. Behav.* **2017**, *177*, 215–220. [CrossRef] [PubMed]
12. Moreira, A.; Bradley, P.; Carling, C.; Arruda, A.; Spigolon, L.M.P.; Franciscon, C.; Aoki, M.S. Effect of a congested match schedule on immune-endocrine responses, technical performance and session-RPE in elite youth soccer players. *J. Sports Sci.* **2016**, *34*, 2255–2261. [CrossRef] [PubMed]
13. Nakamura, D.; Akimoto, T.; Suzuki, S.; Kono, I. Daily changes of salivary secretory immunoglobulin A and appearance of upper respiratory symptoms during physical training. *J. Sports Med. Phys. Fit.* **2006**, *46*, 152–157.
14. Papacosta, E.; Nassis, G.P. Saliva as a tool for monitoring steroid, peptide and immune markers in sport and exercise science. *J. Sci. Med. Sport* **2011**, *14*, 424–434. [CrossRef]
15. Thorpe, R.; Sunderland, C. Muscle Damage, Endocrine, and Immune Marker Response to a Soccer Match. *J. Strength Cond. Res.* **2012**, *26*, 2783–2790. [CrossRef]
16. Rowell, A.E.; Aughey, R.J.; Hopkins, W.G.; Stewart, A.M.; Cormack, S.J. Identification of Sensitive Measures of Recovery after External Load from Football Match Play. *Int. J. Sports Physiol. Perform.* **2017**, *12*, 969–976. [CrossRef]
17. Crewther, B.T.; Lowe, T.; Weatherby, R.P.; Gill, N.; Keogh, J. Neuromuscular Performance of Elite Rugby Union Players and Relationships with Salivary Hormones. *J. Strength Cond. Res.* **2009**, *23*, 2046–2053. [CrossRef]
18. Cormack, S.J.; Newton, R.U.; McGuigan, M.R. Neuromuscular and Endocrine Responses of Elite Players to an Australian Rules Football Match. *Int. J. Sports Physiol. Perform.* **2008**, *3*, 359–374. [CrossRef]

19. Nunes, J.A.; Crewther, B.T.; Ugrinowitsch, C.; Tricoli, V.; Viveiros, L.; de Rose, D.; Aoki, M.S. Salivary Hormone and Immune Responses to Three Resistance Exercise Schemes in Elite Female Athletes. *J. Strength Cond. Res.* **2011**, *25*, 2322–2327. [CrossRef]
20. Morgans, R.; Adams, D.; Mullen, R.; McLellan, C.; Williams, M. An English championship league soccer team did not experience impaired physical match performance in a second match after 75 hours of recovery. *J. Aust. Str. Cond.* **2014**, *22*, 16–23.
21. Filaire, E.; Lac, G.; Pequignot, J.M. Biological, hormonal, and psychological parameters in professional soccer players throughout a competitive season. *Percept. Mot. Ski.* **2003**, *97 Pt 2*, 1061–1072. [CrossRef]
22. Handziski, Z.; Maleska, V.; Petrovska, S.; Nikolik, S.; Mickoska, E.; Dalip, M.; Kostova, E. The changes of ACTH, cortisol, testosterone and testosterone/cortisol ratio in professional soccer players during a competition half-season. *Bratisl. Lekárske Listy* **2006**, *107*, 259.
23. Winter, E.M.; Maughan, R.J. Requirements for ethics approvals. *J. Sports Sci.* **2009**, *27*, 985. [CrossRef] [PubMed]
24. Fredericks, S.; Fitzgerald, L.; Shaw, G.; Holt, D.W. Changes in salivary immunoglobulin A (IgA) following match-play and training among English premiership footballers. *Med. J. Malays.* **2012**, *67*, 155.
25. Venkataraman, S.; Munoz, R.; Candido, C.; Witchel, S.F. The hypothalamic-pituitary-adrenal axis in critical illness. *Endocrinol. Metab. Clin.* **2007**, *8*, 365–373. [CrossRef] [PubMed]
26. Coad, S.; McLellan, C.; Whitehouse, T.; Gray, B. Validity and Reliability of a Novel Salivary Immunoassay for Individual Profiling in Applied Sports Science. *Res. Sports Med.* **2015**, *23*, 140–150. [CrossRef]
27. A Macdonald, L.; Bellinger, P.M.; Minahan, C.L. Reliability of salivary cortisol and immunoglobulin-A measurements from the IPRO®before and after sprint cycling exercise. *J. Sports Med. Phys. Fit.* **2017**, *57*, 1680–1686. [CrossRef]
28. Dunbar, J.; Jehanli, A.; Skelhorn, S. Investigating the use of a point of care SIgA test in the sporting environment. In Proceedings of the 10th International Society of Exercise Immunology, Oxford, UK, 11–13 July 2011; Volume 85.
29. Júnior, O.O.; Chiari, R.; Lopes, W.R.; Abreu, K.C.; Lopes, A.D.; Fialho, G.; Lasmar, R.C.P.; Bittencourt, N.F.N.; Leopoldino, A.A.O. Validation and reliability between external load analysis devices for soccer players. *Rev. Bras. Med. Esporte* **2022**, *28*, 286–290. [CrossRef]
30. Kleiman, E. Data Management Tools for Real-Time Monitoring/Ecological Momentary Assessment Data [R Package EMA-tools version 0.1.3]. 2017. Available online: https://cran.r-project.org/web/packages/EMAtools/index.html (accessed on 25 July 2022).
31. Proctor, G.B.; Carpenter, G.H. Salivary Secretion: Mechanism and Neural Regulation. In *Monographs in Oral Science*; KARGER: Basel, Switzerland, 2014; Volume 24, pp. 14–29, ISBN 0077-0892.
32. Engeland, C.G.; A Bosch, J.; Rohleder, N. Salivary biomarkers in psychoneuroimmunology. *Curr. Opin. Behav. Sci.* **2019**, *28*, 58–65. [CrossRef]
33. Moreira, A.; Freitas, C.G.; Nakamura, F.Y.; Drago, G.; Drago, M.; Aoki, M.S. Effect of Match Importance on Salivary Cortisol and Immunoglobulin A Responses in Elite Young Volleyball Players. *J. Strength Cond. Res.* **2013**, *27*, 202–207. [CrossRef]
34. Bosch, J.A.; de Geus, E.J.; Kelder, A.; Veerman, E.C.; Hoogstraten, J.; Amerongen, A.V. Differential effects of active versus passive coping on secretory immunity. *Psychophysiology* **2001**, *38*, 836–846. [CrossRef] [PubMed]
35. Moreira, A.; McGuigan, M.R.; Arruda, A.F.; Freitas, C.G.; Aoki, M.S. Monitoring Internal Load Parameters during Simulated and Official Basketball Matches. *J. Strength Cond. Res.* **2012**, *26*, 861–866. [CrossRef] [PubMed]
36. Dhabhar, F.S. The short-term stress response—Mother nature's mechanism for enhancing protection and performance under conditions of threat, challenge, and opportunity. *Front. Neuroendocr.* **2018**, *49*, 175–192. [CrossRef] [PubMed]
37. Nicolette, C.B.; Bishop, N.C.; Gleeson, M. Acute and chronic effects of exercise on markers of mucosal immunity. *Front. Biosci.* **2009**, *4*, 4444–4456. [CrossRef]
38. Moreira, A.; Bacurau, R.; Napimoga, M.; Arruda, A.; Freitas, C.; Drago, G.; Aoki, M.S. Salivary IL-21 and IgA responses to a competitive match in elite basketball players. *Biol. Sport* **2013**, *30*, 243–247. [CrossRef]
39. Maya, J.; Marquez, P.; Peñailillo, L.; Contreras-Ferrat, A.; Deldicque, L.; Zbinden-Foncea, H. Salivary Biomarker Responses to Two Final Matches in Women's Professional Football. *J. Sports Sci. Med.* **2016**, *15*, 365–371.
40. Moreira, A.; Arsati, F.; Cury, P.R.; Franciscon, C.; de Oliveira, P.R.; de Araújo, V.C. Salivary Immunoglobulin A Response to a Match in Top-Level Brazilian Soccer Players. *J. Strength Cond. Res.* **2009**, *23*, 1968–1973. [CrossRef]
41. Mortatti, A.L.; de Oliveira, R.S.C.; Pinto, J.C.B.D.L.; Galvão-Coelho, N.L.; de Almeida, R.N.; Aoki, M.S.; Moreira, A. A Congested Match Schedule Alters Internal Match Load and Affects Salivary Immunoglobulin A Concentration in Youth Soccer Players. *J. Strength Cond. Res.* **2020**, *36*, 1655–1659. [CrossRef]
42. Van Paridon, K.N.; A Timmis, M.; Nevison, C.M.; Bristow, M. The anticipatory stress response to sport competition; a systematic review with meta-analysis of cortisol reactivity. *BMJ Open Sport Exerc. Med.* **2017**, *3*, e000261. [CrossRef]
43. Alix-Sy, D.; Le Scanff, C.; Filaire, E. Psychophysiological responses in the pre-competition period in elite soccer players. *J. Sports Sci. Med.* **2008**, *7*, 446–454.
44. Edwards, D.A.; Wetzel, K.; Wyner, D.R. Intercollegiate soccer: Saliva cortisol and testosterone are elevated during competition, and testosterone is related to status and social connectedness with teammates. *Physiol. Behav.* **2006**, *87*, 135–143. [CrossRef] [PubMed]
45. Gozansky, W.S.; Lynn, J.S.; Laudenslager, M.; Kohrt, W.M. Salivary cortisol determined by enzyme immunoassay is preferable to serum total cortisol for assessment of dynamic hypothalamic-pituitary-adrenal axis activity. *Clin. Endocrinol.* **2005**, *63*, 336–341. [CrossRef] [PubMed]

Int. J. Environ. Res. Public Health **2022**, *19*, 11784

46. Arruda, A.F.; Aoki, M.S.; Paludo, A.C.; Moreira, A. Salivary steroid response and competitive anxiety in elite basketball players: Effect of opponent level. *Physiol. Behav.* **2017**, *177*, 291–296. [CrossRef] [PubMed]
47. Slimani, M.; Baker, J.S.; Cheour, F.; Taylor, L.; Bragazzi, N.L. Steroid hormones and psychological responses to soccer matches: Insights from a systematic review and meta-analysis. *PLoS ONE* **2017**, *12*, e0186100. [CrossRef] [PubMed]
48. Augusto, D.; Brito, J.; Aquino, R.; Figueiredo, P.; Eiras, F.; Tannure, M.; Veiga, B.; Vasconcellos, F. Contextual Variables Affect Running Performance in Professional Soccer Players: A Brief Report. *Front. Sports Act. Living* **2021**, *3*, 350. [CrossRef] [PubMed]
49. Chmura, P.; Liu, H.; Andrzejewski, M.; Chmura, J.; Kowalczuk, E.; Rokita, A.; Konefał, M. Is there meaningful influence from situational and environmental factors on the physical and technical activity of elite football players? Evidence from the data of 5 consecutive seasons of the German Bundesliga. *PLoS ONE* **2021**, *16*, e0247771. [CrossRef]
50. Morgans, R.; Orme, P.; Bezuglov, E.; Di Michele, R. Technical and physical performance across five consecutive seasons in elite European Soccer. *Int. J. Sports Sci. Coach* **2022**, 1–9. [CrossRef]
51. Serpell, B.; Horgan, B.G.; Colomer, C.M.; Field, B.; Halson, S.; Cook, C. Sleep and Salivary Testosterone and Cortisol during a Short Preseason Camp: A Study in Professional Rugby Union. *Int. J. Sports Physiol. Perform.* **2019**, *14*, 796–804. [CrossRef]

International Journal of
***Environmental Research
and Public Health***

Article

Effect of Small-Sided Games with and without the Offside Rule on Young Soccer Players: Reliability of Physiological Demands

Igor Junio Oliveira Custódio [†], Renan Dos Santos [†], Rafael de Oliveira Ildefonso, André Andrade, Rodrigo Diniz, Gustavo Peixoto, Sarah Bredt, Gibson Moreira Praça and Mauro Heleno Chagas *

School of Physical Education, Physiotherapy and Occupational Therapy, Federal University of Minas Gerais, Av. Antônio Carlos, 6627, CEP 31270-901, Belo Horizonte 31270-901, Brazil
* Correspondence: mauroufmg@hotmail.com; Tel./Fax: +55-31-3409-7443
† These authors contributed equally to this work.

Abstract: This study aimed to compare the physiological demand between three vs. three small-sided games (SSGs) with (3vs.3$_{WITH}$) and without (3vs.3$_{WITHOUT}$) the offside rule, as well as the within- and between-session reliability of this demand. Twenty-four U-17 soccer athletes performed various three vs. three (plus goalkeepers) SSGs with and without the offside rule. The data collection was performed within an eight-week period. Athletes' heart rate was monitored during the SSG. The variables analyzed were the percentage mean heart rate (HR$_{MEAN\%}$) and the percentage peak heart rate (HR$_{PEAK\%}$). For the analysis of within-session reliability, the mean value of the first two and last two SSG bouts performed within one day were used. The between-session reliability was calculated using the mean value of the four SSG bouts of each SSG type performed on two different days. In both SSGs, the values for reliability were significant and were classified as moderate to excellent. There were no significant differences in the physiological demand among SSG types. We concluded that the offside rule does not influence the physiological demand in a three vs. three SSG and the HR$_{MEAN\%}$ and HR$_{PEAK\%}$ present moderate to excellent reliability in a three vs. three SSG with and without the offside rule.

Keywords: small-sided games; task constraints; physiological demand; young soccer player; peak heart rate; offside rule; reliability

Citation: Custódio, I.J.O.; Dos Santos, R.; de Oliveira Ildefonso, R.; Andrade, A.; Diniz, R.; Peixoto, G.; Bredt, S.; Praça, G.M.; Chagas, M.H. Effect of Small-Sided Games with and without the Offside Rule on Young Soccer Players: Reliability of Physiological Demands. *Int. J. Environ. Res. Public Health* **2022**, *19*, 10544. https://doi.org/10.3390/ijerph191710544

Academic Editor: Paul B. Tchounwou

Received: 18 July 2022
Accepted: 20 August 2022
Published: 24 August 2022

Publisher's Note: MDPI stays neutral with regard to jurisdictional claims in published maps and institutional affiliations.

1. Introduction

In recent years, the physical conditioning of soccer players has developed according to an integrated approach involving tactical and technical aspects of the game [1,2]. In this context, small-sided games (SSGs) provide high-intensity activity, including both tactical and technical demands, and optimize the available training time [3]. Knowledge of the effect of changing SSG characteristics (e.g., the number of players per team, the pitch size, and the rules) helps strength and conditioning coaches to adequately prescribe an SSG during the training process [4]. Although there are many studies on the effect of changing the pitch size and the number of players in a team [5–7], there has been less research on how rule changes in SSGs affect the players' physical and physiological responses [8–11]. One task constraint that can induce changes in players' available space is the offside rule, as it might reduce the effective playing area when the defending team moves towards the opponent's goal. To the best of our knowledge, the influence of the offside rule on the physiological demands of SSGs has not been investigated. Considering the importance of this rule on the game dynamics and the possibility of implementing it during game-based tasks such as an SSG, it is essential to understand its impact on athletes' physiological responses.

The players' movements and displacements during official matches are determined by the effective playing area, which is influenced by the offside rule. This constraint causes the playing area to be dynamic, restricting or allowing players to move across the field length according to the position of teammates and opponents [12]. Hence, the relative area (i.e., area per player) also constantly changes during the game [13]. Previous studies have suggested reducing the relative area by decreasing the absolute pitch size (area in m^2) while maintaining the number of players [14], or keeping the absolute pitch size and increasing the number of players [15]. A smaller relative area generally decreases players' physical [9,16] and physiological [4] responses, because it constrains players' displacements. Therefore, another possibility to modulate the relative area in SSGs is the inclusion of the offside rule, because it can reduce the effective space in which players can move. However, some studies on soccer SSGs have included the offside rule [17,18], while others have not [19,20]. Castillo et al. [9] compared the physical demands of a soccer SSG with and without the offside rule and found a greater total distance and a larger distance covered between 13 and 16 km/h on the pitch without the offside rule. Therefore, it might be expected that non-offside SSGs lead to greater physiological responses from the players. Nonetheless, the influence of this rule on the relative area and consequently on the physical and physiological demands of soccer SSGs requires deeper investigation. Moreover, this knowledge may add a new interpretation to previous studies on SSGs that have or have not implemented the offside rule. Understanding the impact of this rule on athletes' responses can help the coaches to better use game-based activities during training.

Another critical issue regarding the use of SSGs is their reliability as a means of training. This analysis is crucial to test whether specific demands can be achieved when an SSG format is repeatedly applied during the training process. Weir [21] suggested using the intraclass correlation coefficient (ICC) and the standard error of the measurement (SEM) to analyze the reliability. The ICC provides information on the variability between individuals and the consistency of this variability in repeated test measures [21], while the SEM reflects the degree of fluctuation of the individual's scores in a test or condition, representing the expected natural variability (the random error) for that score [21]. Some studies have investigated the reliability of the physiological responses during different soccer SSGs and presented within- [20] and between-session designs [22]. Many of these studies showed high reliability for physiological demands [22–28]. A recent systematic review indicated that internal loads—average heart rate (%HRavg), peak heart rate (%HRpeak), and maximum heart rate (%HRmax)—showed small within-session variations (~0.5–6% of change between the lowest and the highest sets/repetitions), irrespective to the SSG format. Therefore, it is possible to expect high reliability of internal load measures in both with and without offside SSGs in the current study [29].

Considering these issues, this study aimed to (i) compare the physiological demands of a three vs. three SSG with and without the offside rule and (ii) to verify the within- and between-session reliability in these two SSGs.

2. Materials and Methods

2.1. Participants

Twenty-four U-17 male soccer athletes (age: 16.7 ± 0.6 years; body mass: 64.8 ± 6.7 kg; height: 176.5 ± 6.5 cm; body fat: 9.7 ± 1.6%; and estimated VO_{2MAX}: 52.1 ± 2.5 mL·kg^{-1}·min^{-1}) from an elite club participated in this study. This club was considered elite as players compete at the national level regularly. The club achieved first position in the national U-18 competition in the same year the data collection was performed. The athletes competed at a national level and had seven training sessions per week. Data from two athletes were excluded from the analyses due to technical problems, which reduced the final sample to twenty-two players. Players were included if they volunteered to participate in the study and were not injured or returning from injury. On the other hand, the exclusion criteria comprised being injured, not participating in the whole data collection, or refusing to provide written consent to participate in the study. Goalkeepers participated in the data collection but were not evaluated. The participants and their legal guardians were informed about all the research procedures and provided written consent for participating in the study. The local Ethics Committee from the Universidade Federal de Minas Gerais (70103017.0.0000.5149) approved the study, and all the guidelines from the Declaration of Helsinki were followed.

2.2. Teams' Composition for the SSG

The 24 athletes were randomly allocated into eight teams of three players (A to H). Each team had a defender, a midfielder, and a forward to allow teams to explore the physical, technical, and tactical specificities of each playing position during the different SSGs [30,31]. The eight teams were divided into two groups. Group 1 was composed of teams A to D, and Group 2 was composed of teams E to H. Each team within the group played against the same opponent during the entire study (e.g., Team A always played against Team B) to reduce the possible variability related to differences in the opposing teams during the SSGs [32]. The procedures for the composition of the teams and groups are described in Figure 1.

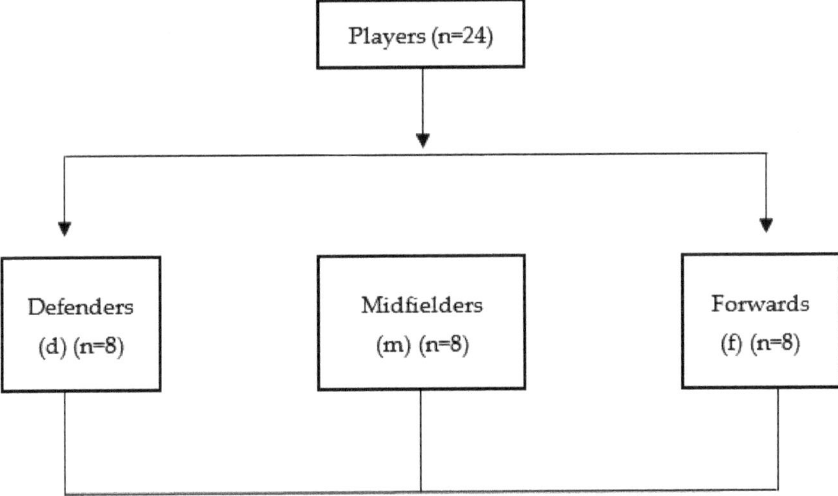

GROUP 1				GROUP 2			
Team	**Position**			**Team**	**Position**		
A	d	m	f	E	d	m	f
B	d	m	f	F	d	m	f
C	d	m	f	G	d	m	f
D	d	m	f	H	d	m	f

Figure 1. Team and group composition procedures. Legend: d = defender; m = midfielder; f = forward.

2.3. Data Collection

Athletes performed several 3 vs. 3 SSGs (where goalkeepers were included but not evaluated) with (3vs.3$_{WITH}$) and without (3vs.3$_{WITHOUT}$) the offside rule. Both of the SSGs were played in the 3 vs. 3 format, on a 36 × 27 m pitch of natural grass, with goals measuring 6 × 2 m (see Figure 2). In the 3vs.3$_{WITH}$ game, two referees were positioned on the sides of the pitch to observe the game and apply the offside rule when necessary. The defending team received a free kick when an offside situation was detected. In the 3vs.3$_{WITHOUT}$ game, the offside rule was not applied, so players could play freely. Each session comprised four SSG bouts, which lasted for four minutes, with five minutes of passive rest. Additional balls were placed around the pitch to ensure a fast game restart when the ball went off the pitch. Coaches and researchers did not give the players verbal encouragement or technical instructions.

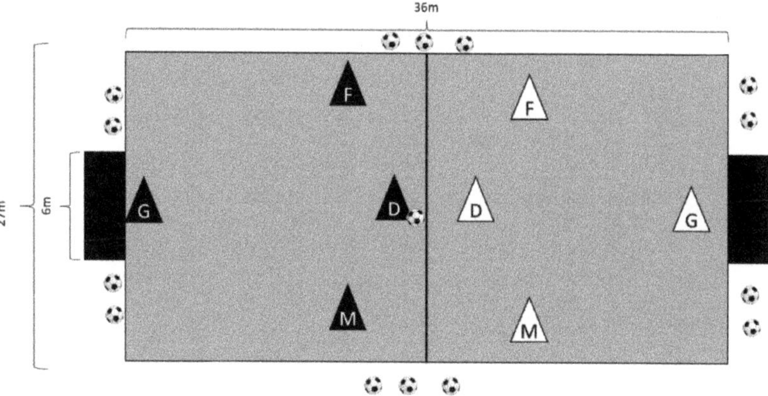

Figure 2. Representation of the 3vs.3$_{WITHOUT}$ game. Legend: G = goalkeeper; D = defender; M = midfielder; F = forward.

The SSGs were performed on Tuesdays and Wednesdays for eight consecutive weeks at the end of the competitive season. We chose the same weekdays to minimize the influence of the distribution of training loads on athletes' physical responses. Group 1 performed the SSGs during the first four weeks (one SSG format each day), while Group 2 performed the SSGs in the last four weeks. This was to avoid a long break between SSG sessions for each team, which could lead to changes in physical fitness, and to minimize the disruption to the athletes' training routines. Therefore, each SSG format was performed twice, with an interval of six to eight days between trials for each SSG format, according to the club availability.

To standardize the influence of circadian rhythm on the observed responses, all sessions were performed at the same time (between 8 a.m. and 10:30 a.m.). The mean (the standard deviation) temperature and relative humidity of all sessions were 31.1 °C (± 2.6 °C) and 28.1% (± 4.8%), respectively, recorded by a portable digital thermometer (Big Digit Hygro-Thermometer, Extech Instruments, Massachusetts, EUA).

To control for the possible effect of changes in physical conditioning on the reliability analysis, athletes performed the Yoyo Intermittent Recovery Test Level 1 (Yo-YoIR1) [31] and a 20 m sprint test one week before and two weeks after the data collection.

In detail, the protocol used for the 20 m sprint test consisted of taking four attempts at the 20 m test, and time recording the distance covered. An interval of three minutes of passive recovery between attempts was established. It is noteworthy that the distance of 20 m was chosen for the measurement of running speed due to evidence that, in official games, sprint running distances longer than 20 m are infrequent [33].

The Yo-YoIR1, on the other hand, is an intermittent, progressive aerobic capacity test, in which athletes perform a series of round-trip runs on a 20 m course [31]. So after each round trip, there is an interval of 10 seconds of active rest in which the athlete trots or walks a course of 10 m, covering 5 m going and 5 m returning. The running speed is determined by sound signals, starting at 10 km/h and increasing progressively throughout the test. In the present study, when the athlete was unable to maintain the rhythm (the speed) determined by the sound signals for two consecutive series, the test was closed, and the total distance covered was recorded. The peak heart rate achieved during Yo-YoIR1 was considered as the athletes' maximum heart rate and was used to relativize heart rate values as a percentage of the maximum.

2.4. Physiological Demand

The heart rate (HR) of the players during the SSGs was recorded using a 1 Hz heart rate monitor (Polar T31 Electro Oy®, Kempele, Finland). The reliability of this device has been previously tested in the literature. Physiological demands were characterized by the percentage of mean heart rate ($HR_{MEAN\%}$) and the percentage of peak heart rate ($HR_{PEAK\%}$). The $HR_{MEAN\%}$ was calculated as the mean of all the values recorded by HR monitors during the SSG bouts (HR values of the rest intervals were excluded). The $HR_{PEAK\%}$ was considered to be the highest value recorded during the SSG bouts. All HR values were relativized by the peak HR presented by each athlete in the Yo-YoIR1.

2.5. Statistical Analyses

The data did not present significant deviations from normality (using Shapiro–Wilk's test) or homoscedasticity (using Levene's test). An independent t-test was used to compare means between the $3vs.3_{WITH}$ and $3vs.3_{WITHOUT}$ games. Cohen's d effect size was calculated to characterize the magnitude of the significant differences in paired comparisons and was classified as insignificant (<0.19), small (0.20–0.49), medium (0.50–0.79), or large ($\geq$0.80) [32].

For the within-session reliability of the $HR_{MEAN\%}$ and $HR_{PEAK\%}$ for the $3vs.3_{WITH}$ and $3vs.3_{WITHOUT}$ games, athletes' mean values of the first two and the last two SSG bouts in each session (day 1 and day 2) were used. To determine the between-session reliability, athletes' mean values of the four SSG bouts performed in each session were used. For both within- and between-session reliability, the intraclass correlation coefficient 2,k ($ICC_{2,k}$) and the standard error of the measurement (SEM) were used [21]. The $ICC_{2,k}$ values were classified as weak (<0.4), moderate (0.40–0.59), good (0.60–0.74), or excellent (0.75–1.00) [34].

A two-way analysis of variance (groups × moments) was used to compare the data on aerobic power (from the Yo-YoIR1) and sprint performance (from the 20 m sprint) among the two groups and moments (from the pre- and post-data collection).

The level of statistical significance was set at 5% (α = 0.05). All analyses were performed using SPSS version 23.0 (Chicago, IL, USA).

3. Results

Table 1 shows the descriptive data (the means and standard deviations) of $HR_{MEAN\%}$ and $HR_{PEAK\%}$ in the investigated SSGs. There were no significant differences between the SSGs with and without the offside rule (giving a small effect size).

Table 1. Means (standard deviations) of the variables related to the physiological demand of SSGs with and without the offside rule.

	$3vs.3_{WITH}$	$3vs.3_{WITHOUT}$			
	Mean (SD)	Mean (SD)	*p*-Value	ES	Interpretation
$HR_{PEAK\%}$	94.8 (2.1)	94.4 (2.1)	0.14	0.33	Small
$HR_{MEAN\%}$	87.4 (2.9)	87.1 (2.3)	0.46	0.16	Insignificant

Legend: $3vs.3_{WITH}$ = small-sided games with the offside rule; $3vs.3_{WITHOUT}$ = small-sided games without the offside rule; $FC_{PEAK\%}$ = percentage peak heart rate; $FC_{MEAN\%}$ = percentage mean heart rate.

Table 2 shows the within-session (bouts within days 1 and 2) intraclass correlation coefficient values (95% CI), the ICC classification, and the SEM values for the variables related to the physiological demand of SSGs with and without the offside rule. The ICC values were classified as "good" or "excellent" (values above 0.60), except for the $HR_{MEAN\%}$, which was classified as "moderate" on day 2.

Table 2. Within-session intraclass correlation coefficients (95% CI), ICC classification, and SEM for the variables related to the physiological demand of SSGs with and without the offside rule.

	$HR_{PEAK\%}$	$HR_{MEAN\%}$	$HR_{PEAK\%}$	$HR_{MEAN\%}$
	3vs.3$_{WITH}$—DAY 1		3vs.3$_{WITH}$—DAY 2	
ICC (95% CI)	0.76 * (0.36–0.91)	0.85 * (0.44–0.95)	0.75 * (0.12–0.91)	0.73 * (0.32–0.89)
ICC Classification	Excellent	Excellent	Excellent	Good
SEM (%)	1.3	1.7	1.4	2.1
	3vs.3$_{WITHOUT}$—DAY 1		3vs.3$_{WITHOUT}$—DAY 2	
ICC (95% CI)	0.61 * (−0.23–0.87)	0.62 * (−0.21–0.87)	0.73 * (−0.20–0.93)	0.58 * (−0.21–0.87)
ICC Classification	Good	Good	Good	Moderate
SEM (%)	1.3	1.7	1.0	1.3

3vs.3$_{WITH}$ = small-sided games with the offside rule; 3vs.3$_{WITHOUT}$ = small-sided games without the offside rule; $FC_{PEAK\%}$ = percentage peak heart rate; $FC_{MEAN\%}$ = percentage mean heart rate; **CI** = confidence interval; **SEM** = standard error of the measurement. * indicates statistical significance ($p < 0.05$).

Table 3 shows the between-session (between days 1 and 2) intraclass correlation coefficient values (95% CI), the ICC classification, and the SEM values for the variables related to the physiological demand of SSGs with and without the offside rule. The ICC values were classified as "good" or "excellent" (values above 0.60), except for the $HR_{MEAN\%}$ in the 3vs.3$_{WITH}$ game, which was classified as "moderate".

Table 3. Between-session intraclass correlation coefficients (95% CI), ICC classification, and SEM for the variables related to the physiological demand of SSGs with and without the offside rule.

	3vs.3$_{WITH}$		3vs.3$_{WITHOUT}$	
	$HR_{PEAK\%}$	$HR_{MEAN\%}$	$HR_{PEAK\%}$	$HR_{MEAN\%}$
ICC (95% CI)	0.62 * (0.09–0.85)	0.56 * (−0.04–0.82)	0.77 * (0.42–0.91)	0.69 * (0.25–0.88)
ICC Classification	Good	Moderate	Excellent	Good
SEM (%)	1.8	2.6	1.4	1.8

3vs.3$_{WITH}$ = small-sided games with the offside rule; 3vs.3$_{WITHOUT}$ = small-sided games without the offside rule; $FC_{PEAK\%}$ = percentage peak heart rate; $FC_{MEAN\%}$ = percentage mean heart rate; **CI** = confidence interval; **SEM** = standard error of the measurement. * indicates statistical significance ($p < 0.05$).

The two way analysis of variance of the control variables (aerobic power—pre-test: 1850.9 ± 288.7 m; post-test: 1950.0 ± 277.6 m and 20 m sprint performance—pre-test: 22.7 ± 0.6 km/h; post-test: 23.3 ± 0.6 km/h) showed no significant interaction (aerobic power—F = 0.68; $p = 0.41$; 20-m sprint performance—F = 0.985; $p = 0.325$) or main effects (aerobic power—F = 3.47; $p = 0.07$; 20 m sprint performance—F = 0.352; $p = 0.556$). These data show the lack of differences in physical conditioning during the period of the data collection, mitigating the possible effect of variability on the between-session reliability analysis.

4. Discussion

This study aimed to investigate the effect of the offside rule on the physiological demands of three vs. three soccer SSGs in U-17s and the reliability of the physiological demands in three vs. three SSGs with and without the offside rule. The results show that the physiological demands, characterized by the $HR_{PEAK\%}$ and $HR_{MEAN\%}$, did not differ among the SSGs with and without the offside rule, and thus, our hypothesis was rejected. Furthermore, the within and between-session reliability of physiological demands confirmed our hypothesis, with moderate to excellent ICC values for all variables, regardless of the rules of the SSGs.

We expected that the offside rule would decrease the physiological demands of the SSGs because of the reduction in the effective playing area. This hypothesis was based on previous results that showed a decrease in the physiological demands when the absolute playing area was decreased for the same number of players or when the number of players was increased within the same playing area [16]. These changes result in smaller relative areas (i.e., area per player), restricting the available space for players to move around and, consequently, reducing the intensity of the game (i.e., lower physiological demands) [14,35]. A previous systematic review included studies with similar playing areas and showed that reducing the relative area per player tends to reduce physical and physiological responses [16]. However, the results of the present study do not corroborate this hypothesis. A possible explanation for these divergent results may be related to the magnitude of the change in the effective playing area in the three vs. three SSG with the offside rule. Previous studies have shown that small changes in the relative area may not be sufficient to influence players' physical [9] and physiological [4] responses. Specifically, the reduction in the effective playing area depends on the defending team moving up the pitch to constrain the available space for the offensive team. Therefore, the number of times the players adopted this behavior might have been smaller than what was required to induce different responses when considering the whole bout. Moreover, although the heart rate has been widely used in studies on SSGs [36] and is considered a valid variable to measure SSG intensity [37], it may not be sensitive enough to detect differences in the frequency of specific actions (i.e., jumps, duels, accelerations, decelerations, sprints, and changes of direction) during the game, which could, in turn, also reflect game intensity [38]. Considering this issue, future studies should collect information through other variables, such as accelerations, decelerations, mean speed, and distances covered in different speed zones, to increase the understanding of exercise intensity during game-based activities, such as SSGs [39].

The $HR_{PEAK\%}$ and $HR_{MEAN\%}$ values found in both SSGs investigated in this study are similar to those reported in previous studies on the three vs. three SSG format performed by soccer players of a similar age (Sub-17) [40–42]. Furthermore, studies on SSG training (training periods above four weeks) indicate the necessity for HR_{mean} values to be above 80% of HR_{max} to improve aerobic performance [43–46]. Therefore, the results of the present study reinforce the potential use of different SSGs for the improvement of aerobic performance in soccer athletes, including the offside rule.

The investigation of SSG reliability is essential to support using SSGs during training. In addition, with the knowledge of the demands imposed on athletes by different SSGs, strength and conditioning coaches can examine if those demands are reproducible when the same SSG is performed at different moments. In the present study, high ICC (>0.60) and low SEM (<1.7) values were found in the within-session reliability analysis of HR_{MEAN}. These data corroborate the results of previous studies on the reliability of heart rate variables collected during SSGs, despite the differences in the SSG formats. Hill-Haas et al. [24] compared different SSG formats (two vs. two, four vs. four, and six vs. six) and found percentage values of SEM (SEM%) of 1.9 and 4.4% for the $HR_{PEAK\%}$ and 1.1 and 3.6% for the $HR_{MEAN\%}$. Another study also reported small SEM percentage values for the $HR_{MEAN\%}$ (5.4%) and $HR_{PEAK\%}$ (3.0%) in a three vs. three SSG with similar characteristics [22]. Finally, Stevens et al. [28] found good reliability values for the $HR_{MEAN\%}$ during a six vs. six SSG (ICC = 0.61 and SEM% = 2.2%). On the other hand, the results of the present study on between-session reliability suggest good reproducibility of the $HR_{PEAK\%}$ and $HR_{MEAN\%}$, despite an interval of one week between the sessions (ICC = 0.56), with a low variability among these measures (SEM < 2.6%). These results are similar to previous research that indicated good reproducibility for the physiological demands represented by heart rate variables in different SSGs. Da Silva et al. [23] and Rampinini et al. [27] investigated the reliability of the $HR_{MEAN\%}$ in SSGs with different numbers of players and pitch sizes and found that values of SEM percentage ranged from 2.2 and 3.4%, and the percentage of typical error (TE%) values (similar to SEM) ranged from 2.0% and 5.4%, respectively. Additionally, Hill-Haas et al. [25] found low variability for the $HR_{MEAN\%}$, with TE% values ranging from 2 and 4% in a four vs. four SSG. This result is similar to that found by Hulka et al. [47], which showed high ICC (0.88) and low SEM% (2.35%) values in a four vs. four SSG. Additionally, both with and without the offside rule, SSGs showed similar classifications regarding the reliability measures. However, when looking at both within- and between-session reliability, the SSG without the offside rule showed lower ICC values than the SSG with it. It has been proposed in the literature that a higher movement variability can be detected in lesser-known game formats [48–50]. It can be argued that U-17 soccer players usually engage in more specific tasks than those that are general game-based tasks—therefore, the game with the offside rule seems to be more representative of the requirements of the official match. Consequently, the reduction in the reliability might indicate a more variable displacement behavior in the SSGwithout condition due to the players' need to readapt to the new constraints.

This study investigated U-17 athletes, which hinders the generalization of the results to other age categories. Future studies should be carried out with athletes of different ages to provide more precise information on the physiological demands of the three vs. three SSGs investigated in this study. Moreover, this study did not monitor athletes' recovery levels during the data collection, which could have added a deeper understanding of athletes' conditions while recording the variables. In this case, further research should investigate athletes' recovery behavior over SSG bouts and between training sessions to provide information that better supports the use of SSGs for the physical conditioning of soccer players.

5. Conclusions

Using the offside rule in a three vs. three SSG did not influence the physiological responses of young soccer athletes. The within- and between-session reliability values of the physiological variables in both SSGs with and without the offside rule were high, supporting the reproducibility of the physiological demands of SSGs despite their natural unpredictability and variability. The absence of difference between the protocols indicates that coaches might choose between the two SSG formats based on other goals—for example, tactical missions related to enlarging the surface area—instead of considering the impact the offside rule will have on players' physiological responses.

Author Contributions: Conceptualization, R.D.S., G.M.P. and M.H.C.; formal analysis, A.A. and R.D.; investigation, R.D.S., I.J.O.C., R.d.O.I. and S.B.; methodology, R.D., G.P. and S.B.; writing—original draft preparation, I.J.O.C., R.d.O.I. and G.P.; writing—review and editing, R.D.S., A.A., S.B. and M.H.C.; project administration, R.D.S. and M.H.C. All authors have read and agreed to the published version of the manuscript.

Funding: This work was supported by the Pró-Reitoria de Pesquisa da Universidade Federal de Minas Gerais (PRPq-UFMG), Coordenação de Aperfeiçoamento de Pessoal de Nível Superior—Brasil (CAPES), Conselho Nacional de Desenvolvimento Científico e Técnológico (CNPq), and Fundação de Amparo à Pesquisa do Estado de Minas Gerais (FAPEMIG).

Institutional Review Board Statement: The study was conducted according to the guidelines of the Declaration of Helsinki, and approved by the Institutional Review Board (or Ethics Committee) of Federal University of Minas Gerais/Brazil (Protocol number 70103017.0.0000.5149 and date of approval was 8 September 2017).

Informed Consent Statement: Informed consent was obtained from all participants involved in the study.

Data Availability Statement: Not applicable.

Conflicts of Interest: The authors declared no potential conflicts of interest with respect to the research, authorship, and/or publication of this article.

References

1. Moran, J.; Blagrove, R.C.; Drury, B.; Fernandes, J.F.; Paxton, K.; Chaabene, H.; Ramirez-Campillo, R. Effects of small-sided games vs. conventional endurance training on endurance performance in male youth soccer players: A meta-analytical comparison. *Sports Med.* **2019**, *49*, 731–742. [CrossRef] [PubMed]
2. Morgans, R.; Orme, P.; Anderson, L.; Drust, B. Principles and practices of training for soccer. *J. Sport Health Sci.* **2014**, *3*, 251–257. [CrossRef]
3. Davids, K.; Araújo, D.; Correia, V.; Vilar, L. How small-sided and conditioned games enhance acquisition of movement and decision-making skills. *Exerc. Sport Sci. Rev.* **2013**, *41*, 154–161. [CrossRef] [PubMed]
4. Sarmento, H.; Clemente, F.M.; Harper, L.D.; Costa, I.T.; Owen, A.; Figueiredo, A.J. Small sided games in soccer–a systematic review. *Int. J. Perform. Anal. Sport* **2018**, *18*, 693–749. [CrossRef]
5. Castellano, J.; Casamichana, D.; Dellal, A. Influence of game format and number of players on heart rate responses and physical demands in small-sided soccer games. *J. Strength Cond. Res.* **2013**, *27*, 1295–1303. [CrossRef]
6. Hodgson, C.; Akenhead, R.; Thomas, K. Time-motion analysis of acceleration demands of 4v4 small-sided soccer games played on different pitch sizes. *Hum. Mov. Sci.* **2014**, *33*, 25–32. [CrossRef]
7. Sangnier, S.; Cotte, T.; Brachet, O.; Coquart, J.; Tourny, C. Planning Training Workload in Football Using Small-Sided Games' Density. *J. Strength Cond. Res.* **2018**, *33*, 2801–2811. [CrossRef]
8. Casamichana, D.; San Román-Quintana, J.; Castellano, J.; Calleja-González, J. Influence of the type of marking and the number of players on physiological and physical demands during sided games in soccer. *J. Hum. Kinet.* **2015**, *47*, 259–268. [CrossRef]
9. Castillo, D.; Raya-González, J.; Manuel Clemente, F.; Yanci, J. The influence of offside rule and pitch sizes on the youth soccer players' small-sided games external loads. *Res. Sports Med.* **2020**, *28*, 324–338. [CrossRef]
10. Dellal, A.; Lago-Penas, C.; Wong, D.P.; Chamari, K. Effect of the number of ball contacts within bouts of 4 vs. 4 small-sided soccer games. *Int. J. Sports Physiol. Perform.* **2011**, *6*, 322–333. [CrossRef]
11. San Román-Quintana, J.; Casamichana, D.; Castellano, J.; Calleja-González, J.; Jukić, I.; Ostojić, S.M. The influence of ball-touches number on physical and physiological demands of large-sided games. *Kinesiol. Int. J. Fundam. Appl. Kinesiol.* **2013**, *45*, 171.
12. Praça, G.M.; Chagas, M.H.; Bredt, S.G.; Andrade, A.G.; Custódio, I.J.; Rochael, M. The influence of the offside rule on players' positional dynamics in soccer small-sided games. *Sci. Med. Footb.* **2021**, *5*, 144–149. [CrossRef]
13. Fradua, L.; Zubillaga, A.; Caro, Ó.; Iván Fernández-García, Á.; Ruiz-Ruiz, C.; Tenga, A. Designing small-sided games for training tactical aspects in soccer: Extrapolating pitch sizes from full-size professional matches. *J. Sports Sci.* **2013**, *31*, 573–581. [CrossRef]
14. Vilar, L.; Duarte, R.; Silva, P.; Chow, J.Y.; Davids, K. The influence of pitch dimensions on performance during small-sided and conditioned soccer games. *J. Sports Sci.* **2014**, *32*, 1751–1759. [CrossRef]
15. Chung, D.; Carvalho, T.; Casanova, F.; Silva, P. Number of players manipulation effect on space and concentration principles of the game representativeness during football small-sided and conditioned games. *J. Phys. Educ. Sport* **2019**, *19*, 381–386. [CrossRef]
16. Praça, G.M.; Chagas, M.H.; Bredt, S.G.; Andrade, A.G. Small-Sided Soccer Games with Larger Relative Areas Result in Higher Physical and Physiological Responses: A Systematic and Meta-Analytical Review. *J. Hum. Kinet.* **2022**, *81*, 163–176. [CrossRef]
17. Hill-Haas, S.V.; Coutts, A.J.; Dawson, B.T.; Rowsell, G.J. Time-motion characteristics and physiological responses of small-sided games in elite youth players: The influence of player number and rule changes. *J. Strength Cond. Res.* **2010**, *24*, 2149–2156. [CrossRef]
18. Praça, G.; Clemente, F.M.; Andrade, A.G.; Morales, J.C.; Greco, P.J. Network analysis in small-sided and conditioned soccer games: The influence of additional players and playing position. *Kinesiol. Int. J. Fundam. Appl. Kinesiol.* **2017**, *49*, 185–193. [CrossRef]
19. Gaudino, P.; Alberti, G.; Iaia, F.M. Estimated metabolic and mechanical demands during different small-sided games in elite soccer players. *Hum. Mov. Sci.* **2014**, *36*, 123–133. [CrossRef]
20. Travassos, B.; Gonçalves, B.; Marcelino, R.; Monteiro, R.; Sampaio, J. How perceiving additional targets modifies teams' tactical behavior during football small-sided games. *Hum. Mov. Sci.* **2014**, *38*, 241–250. [CrossRef]
21. Weir, J.P. Quantifying test-retest reliability using the intraclass correlation coefficient and the SEM. *J. Strength Cond. Res.* **2005**, *19*, 231–240.
22. Bredt, S.G.; Praça, G.M.; Figueiredo, L.S.; Paula, L.V.; Silva, P.C.; Andrade, A.G.; Greco, P.J.; Chagas, M.H. Reliability of physical, physiological and tactical measures in small-sided soccer Games with numerical equality and numerical superiority. *Rev. Bras. Cineantropometria Desempenho Hum.* **2016**, *18*, 602–610. [CrossRef]
23. Silva, C.D.; Impellizzeri, F.M.; Natali, A.J.; de Lima, J.R.; Bara-Filho, M.G.; Silami-Garçia, E.; Marins, J.C. Exercise intensity and technical demands of small-sided games in young Brazilian soccer players: Effect of number of players, maturation, and reliability. *J. Strength Cond. Res.* **2011**, *25*, 2746–2751. [CrossRef]

24. Hill-Haas, S.; Coutts, A.; Rowsell, G.; Dawson, B. Variability of acute physiological responses and performance profiles of youth soccer players in small-sided games. *J. Sci. Med. Sport* **2008**, *11*, 487–490. [CrossRef]
25. Hill-Haas, S.; Rowsell, G.; Coutts, A.; Dawson, B. The reproducibility of physiological responses and performance profiles of youth soccer players in small-sided games. *Int. J. Sports Physiol. Perform.* **2008**, *3*, 393–396. [CrossRef]
26. Ngo, J.K.; Tsui, M.-C.; Smith, A.W.; Carling, C.; Chan, G.-S.; Wong, D.P. The effects of man-marking on work intensity in small-sided soccer games. *J. Sports Sci. Med.* **2012**, *11*, 109.
27. Rampinini, E.; Impellizzeri, F.M.; Castagna, C.; Abt, G.; Chamari, K.; Sassi, A.; Marcora, S.M. Factors influencing physiological responses to small-sided soccer games. *J. Sports Sci.* **2007**, *25*, 659–666. [CrossRef] [PubMed]
28. Stevens, T.G.; De Ruiter, C.J.; Beek, P.J.; Savelsbergh, G.J. Validity and reliability of 6-a-side small-sided game locomotor performance in assessing physical fitness in football players. *J. Sports Sci.* **2016**, *34*, 527–534. [CrossRef] [PubMed]
29. Clemente, F.M.; Aquino, R.; Praça, G.; Rico-González, M.; Oliveira, R.; Silva, A.F.; Sarmento, H.; Afonso, J. Variability of internal and external loads and technical/tactical outcomes during small-sided soccer games: A systematic review. *Biol. Sport* **2021**, *3*, 647–672.
30. Hopkins, W.G. Measures of reliability in sports medicine and science. *Sports Med.* **2000**, *30*, 1–15. [CrossRef] [PubMed]
31. Krustrup, P.; Mohr, M.; Amstrup, T.; Rysgaard, T.; Johansen, J.; Steensberg, A.; Pedersen, P.K.; Bangsbo, J. The yo-yo intermittent recovery test: Physiological response, reliability, and validity. *Med. Sci. Sports Exerc.* **2003**, *35*, 697–705. [CrossRef]
32. Cohen, J. *Statistical Power Analysis for the Behavioral Sciences*; Academic Press: Nova Iorque, NY, USA, 1988; Volume 10, pp. 50012–50018. [CrossRef]
33. Di Salvo, V.; Baron, R.; Tschan, H.; Montero, F.C.; Bachl, N.; Pigozzi, F. Performance characteristics according to playing position in elite soccer. *Int. J. Sports Med.* **2007**, *28*, 222–227. [CrossRef]
34. Cicchetti, D.V. Guidelines, criteria, and rules of thumb for evaluating normed and standardized assessment instruments in psychology. *Psychol. Assess.* **1994**, *6*, 284. [CrossRef]
35. Frencken, W.; Van Der Plaats, J.; Visscher, C.; Lemmink, K. Size matters: Pitch dimensions constrain interactive team behaviour in soccer. *J. Syst. Sci. Complex.* **2013**, *26*, 85–93. [CrossRef]
36. Hill-Haas, S.V.; Dawson, B.; Impellizzeri, F.M.; Coutts, A.J. Physiology of small-sided games training in football. *Sports Med.* **2011**, *41*, 199–220. [CrossRef]
37. Hoff, J.; Wisløff, U.; Engen, L.C.; Kemi, O.J.; Helgerud, J. Soccer specific aerobic endurance training. *Br. J. Sports Med.* **2002**, *36*, 218–221. [CrossRef]
38. Casamichana, D.; Castellano, J.; Calleja-Gonzalez, J.; San Román, J.; Castagna, C. Relationship between indicators of training load in soccer players. *J. Strength Cond. Res.* **2013**, *27*, 369–374. [CrossRef]
39. Bartlett, J.D.; O'Connor, F.; Pitchford, N.; Torres-Ronda, L.; Robertson, S.J. Relationships between internal and external training load in team-sport athletes: Evidence for an individualized approach. *Int. J. Sports Physiol. Perform.* **2017**, *12*, 230–234. [CrossRef]
40. Köklü, Y.; Sert, Ö.; Alemdaroglu, U.; Arslan, Y. Comparison of the physiological responses and time-motion characteristics of young soccer players in small-sided games: The effect of goalkeeper. *J. Strength Cond. Res.* **2015**, *29*, 964–971. [CrossRef]
41. Köklü, Y.; Alemdaroğlu, U. Comparison of the Heart Rate and Blood Lactate Responses of Different Small Sided Games in Young Soccer Players. *Sports* **2016**, *4*, 48. [CrossRef]
42. Aşçı, A. Heart rate responses during small sided games and official match-play in soccer. *Sports* **2016**, *4*, 31. [CrossRef]
43. Impellizzeri, F.M.; Marcora, S.M.; Castagna, C.; Reilly, T.; Sassi, A.; Iaia, F.; Rampinini, E. Physiological and performance effects of generic versus specific aerobic training in soccer players. *Int. J. Sports Med.* **2006**, *27*, 483–492. [CrossRef]
44. Faude, O.; Steffen, A.; Kellmann, M.; Meyer, T. The effect of short-term interval training during the competitive season on physical fitness and signs of fatigue: A crossover trial in high-level youth football players. *Int. J. Sports Physiol. Perform.* **2014**, *9*, 936–944. [CrossRef]
45. Owen, A.L.; Wong, D.P.; Paul, D.; Dellal, A. Effects of a periodized small-sided game training intervention on physical performance in elite professional soccer. *J. Strength Cond. Res.* **2012**, *26*, 2748–2754. [CrossRef]
46. Paul, D.J.; Marques, J.B.; Nassis, G.P. The effect of a concentrated period of soccer specific fitness training with small-sided games on physical fitness in youth players. *J. Sports Med. Phys. Fit.* **2018**, *58*, 962–968. [CrossRef]
47. Hůlka, K.; Weisser, R.; Bělka, J.; Háp, P. Stability of internal response and external load during 4-a-side football game in an indoor environment. *Acta Gymnica* **2015**, *45*, 21–25. [CrossRef]
48. Praça, G.M.; Andrade, A.G.; Bredt, S.G.; Moura, F.A.; Moreira, P.E. Progression to the target vs. regular rules in Soccer small-sided Games. *Sci. Med. Footb.* **2022**, *6*, 66–71. [CrossRef]
49. Coutinho, D.; Gonçalves, B.; Santos, S.; Travassos, B.; Wong, D.P.; Sampaio, J. Effects of the pitch configuration design on players' physical performance and movement behaviour during soccer small-sided games. *Res. Sports Med.* **2019**, *27*, 298–313. [CrossRef]
50. Santos, S.; Coutinho, D.; Gonçalves, B.; Abade, E.; Pasquarelli, B.; Sampaio, J. Effects of manipulating ball type on youth footballers' performance during small-sided games. *Int. J. Sports Sci. Coach.* **2020**, *15*, 170–183. [CrossRef]

International Journal of
*Environmental Research
and Public Health*

Article

Perceived Training of Junior Speed Skaters versus the Coach's Intention: Does a Mismatch Relate to Perceived Stress and Recovery?

Ruby T. A. Otter [1,2,*], Anna C. Bakker [3], Stephan van der Zwaard [4,5], Tynke Toering [1], Jos F. A. Goudsmit [6,7], Inge K. Stoter [8] and Johan de Jong [1,3]

1 School of Sports Studies, Hanze University of Applied Sciences, 9747 AS Groningen, The Netherlands
2 Section Anatomie & Medical Physiology, Department Biomedical Sciences of Cells & Systems, University Medical Center Groningen, University of Groningen, 9713 GZ Groningen, The Netherlands
3 Department Human Movement Sciences, University Medical Center Groningen, University of Groningen, 9713 GZ Groningen, The Netherlands
4 Department of Human Movement Sciences, Vrije Universiteit Amsterdam, 1081 HV Amsterdam, The Netherlands
5 Leiden Institute of Advanced Computer Science, Universiteit Leiden, 2300 RA Leiden, The Netherlands
6 School of Sport Studies, Fontys University of Applied Science, 5612 MA Eindhoven, The Netherlands
7 Department of Industrial Design, Eindhoven University of Technology, 5600 MB Eindhoven, The Netherlands
8 Innovatielab Thialf, 8443 DA Heerenveen, The Netherlands
* Correspondence: t.a.otter@pl.hanze.nl

Abstract: The aim of this observational study was to examine the differences between training variables as intended by coaches and perceived by junior speed skaters and to explore how these relate to changes in stress and recovery. During a 4-week preparatory period, intended and perceived training intensity (RPE) and duration (min) were monitored for 2 coaches and their 23 speed skaters, respectively. The training load was calculated by multiplying RPE by duration. Changes in perceived stress and recovery were measured using RESTQ-sport questionnaires before and after 4 weeks. Results included 438 intended training sessions and 378 executed sessions of 14 speed skaters. A moderately higher intended (52:37 h) versus perceived duration (45:16 h) was found, as skaters performed fewer training sessions than anticipated (four sessions). Perceived training load was lower than intended for speed skating sessions (-532 ± 545 AU) and strength sessions (-1276 ± 530 AU) due to lower RPE scores for skating (-0.6 ± 0.7) or shorter and fewer training sessions for strength ($-04:13 \pm 02:06$ hh:mm). All training and RESTQ-sport parameters showed large inter-individual variations. Differences between intended–perceived training variables showed large positive correlations with changes in RESTQ-sport, i.e., for the subscale's success ($r = 0.568$), physical recovery ($r = 0.575$), self-regulation ($r = 0.598$), and personal accomplishment ($r = 0.589$). To conclude, speed skaters that approach or exceed the coach's intended training variables demonstrated an increased perception of success, physical recovery, self-regulation, and personal accomplishment.

Keywords: training load; adolescent athletes; perceived stress and recovery; monitoring; coaching

Citation: Otter, R.T.A.; Bakker, A.C.; van der Zwaard, S.; Toering, T.; Goudsmit, J.F.A.; Stoter, I.K.; de Jong, J. Perceived Training of Junior Speed Skaters versus the Coach's Intention: Does a Mismatch Relate to Perceived Stress and Recovery? *Int. J. Environ. Res. Public Health* **2022**, *19*, 11221. https://doi.org/10.3390/ijerph191811221

Academic Editors: Rafael Oliveira and João Paulo Brito

Received: 25 July 2022
Accepted: 31 August 2022
Published: 7 September 2022

Publisher's Note: MDPI stays neutral with regard to jurisdictional claims in published maps and institutional affiliations.

1. Introduction

Speed skating is a competitive sport in which a multidisciplinary training program is required for the athletes to perform well at the elite level. Training programs for speed skaters include aerobic-, anaerobic-, and strength training on ice, on the bike, and in the gym [1]. This emphasizes the need for speed skating coaches to find a good balance in their training programs so that the training volume and intensity can be managed by the athletes. The Dutch Speed Skating Association selects speed skaters from the age of 17 years and provides them with a high-quality training program [2]. For many speed skaters, this is the time during which they rapidly increase their training volume to be ready for the senior level, which starts after the age of 19 years. The increasing training volume, in addition to the fact that junior athletes develop in many domains other than sport as well (e.g., school, friendships, etc.), may make each individual vulnerable to non-functional overreaching, overtraining [3], and overuse injuries [4]. On the other hand, too few training loads can lead to non-optimal performance increases. Nearly thirty percent of young athletes experience non-functional overreaching or overtraining at least once in their careers [5]. Therefore, it is important to prescribe a tailored training load for individual athletes.

Coaches plan training sessions based on their point of view of training intensity distribution, taking the frequency, intensity, and duration of sessions into account [6]. Usually, the intended training load (defined as the sum of all training sessions' intensities multiplied by the duration [7]) is prescribed as an external load such as speed, pace, duration, or power. However, the internal training load response of the athlete determines the outcome of the training [8]. The internal load (psychophysiological response) is in turn dependent on the actual external load and personal circumstances of the athletes, such as age [9], experience [10], and perceived stressors [11].

A reliable, valid, and also easy way to monitor internal training load is by logging session Ratings of Perceived Exertion (sRPE) and duration for each training session [12]. This method has shown to be useful for a wide variety of training modalities, such as endurance, interval, and strength training [13]. It can be used for both the prescribed training sessions (intended load) by the coach and the perception of actual training sessions (perceived load) by the athlete so that a comparison can be made [14].

Many studies have shown that there are differences between the intended training load by the coach and the perceived training load by the athletes, using measurements of sRPE and duration. In an attempt to specify the differences, the training sessions have been divided into easy, moderate, and hard. On the RPE CR10 scale [12], easy, moderate, and hard were classified as an RPE < 3, 3–5, and >5, respectively [14–17]. However, studies using the 6 to 20 Borg scale [18] have defined two different classifications for easy, moderate, and hard, i.e., RPE < 11, 11–14, and >14, respectively [19], and RPE < 13, 13–14, and >14, respectively [20]. These differences in classifications lead to differences in interpretation of the results, which makes it hard to compare different studies. In order to have a rough overview of previous studies, a short summary of the results in cyclic sports is given.

Previous studies using the RPE CR10 scale showed no differences in intensity, duration, and training load on average [14–17]. However, the training load of easy sessions was perceived as higher by the athletes. Intended moderate sessions did not show any differences between coach and athlete, except for male cross-country runners who perceive it as harder [17]. Moreover, the training load of intended hard sessions was perceived as lower by runners [14], swimmers [16], and female cross-country runners [17]. For cyclists [15] and male cross-country runners [17], there were no differences for intended hard sessions. Using the Borg scale, it was shown that semi-professional cyclists perceived both intended easy sessions and hard sessions as easier [19].

Overall, it seems that there is quite some variation between the perception of sessions that are meant to be easy, moderate, and hard, while the total training duration and training load are mainly similar. One reason for the variation may be that both adolescent and adult athletes were included in the studies. It has been shown that, during adolescence, the correlation between RPE and heart rate [9], and between the coach's and athlete's RPE, increases with age [10]. This suggests that as adolescent athletes reach adulthood, they may vary less in their appraisal of exertion. Furthermore, there seems to be a difference in several sports modalities such as running, cycling, and swimming. Given that the division of the scales can have a great influence on the results, it is important to visualize the entire width of the scale that has been used to gain more detailed information.

Kenttä and Hassmén's (1998) conceptual model shows that the accumulation of physical and psycho-social stress and/or a shortage of recovery may lead to maladaptations such as decreased performance and an increased injury risk [21]. A consensus statement reads that the outcome of training stimuli is, among other factors, influenced by burdening psycho-social factors [3]. In addition, a recent review described that increased perception of stress is one of the main indicators of functional overreaching [11]. Therefore, regular measurement of perceived stress and recovery can be useful for athletes to prevent maladaptation to training.

Intended training load by the coach and perceived training load by junior speed skaters, and what that means for perceived stress and recovery, have not been investigated. Therefore, the goal of this study is two-fold: (1) to examine the (mis)match between the intended training intensity, duration, and load by the coach and the perceived training intensity, duration, and load by junior speed skaters over the entire range of intensities and for the different modalities of training (skating, cycling, strength training, and other); and (2) to explore how the (mis)match in training variables relates to changes in the perception of stress and recovery by the speed skaters in a four-week preparatory period.

2. Materials and Methods

2.1. Subjects

In this study, the whole regional talent team including 23 junior speed skaters (14 males, 9 females; age, 18 ± 1 y (range: 16–19 y; body mass, 68 ± 7 kg; height, 178 ± 8 cm) and their two coaches participated. The speed skaters were part of one training group divided over two coaches. They all performed at a national and international level with an average personal best 1500 m time 12% above the WR. In speed skating, this equals the sub-elite level [22]. All speed skaters and coaches were informed about the procedures of the study and signed a written informed consent form. This study was approved by the ethics committee of the Department of Human Movement Sciences, University of Groningen, and was in accordance with the Declaration of Helsinki.

2.2. Design

In this observational study, the group of 23 junior speed skaters was monitored over a four-week period. The four weeks were during the general preparation period (June–August 2021), in which the speed skaters executed different training modalities, i.e., speed skating, cycling, strength training, and other training modalities (such as plyometric training and roller skating). For every training session, the intended training as planned by the coach for each individual athlete and the actual training as perceived by the athlete were monitored. All training modalities were included in this study.

2.3. Training Load

During the four weeks, the coaches provided each individual speed skater with a weekly training schedule, which was shared in advance. The schedule included the number of sessions, training modality, a description of the type of training, intensity, the duration of specific parts of the training, and the total intended duration. To define the intended training intensity for each training session, the coaches also filled in the intended Rating of Perceived Exertion (RPE) for each session in a separate spreadsheet. The intended RPE was not shared with the speed skaters. To determine the perceived training load by the speed skaters, the Smartabase athlete management platform was used (version 6.11.6). In this platform, speed skaters filled in the realized training modality, the duration, and the perceived RPE for each training session. The athletes were instructed to do this approximately 30 min after each training session [12]. For the RPE, a modified Borg CR10 scale from 0 (rest) to 10 (maximum) was used [12]. Before the four-week period, all speed skaters and coaches were familiarized with the modified Borg CR10 scale. Intended and perceived training load were calculated using the following formulas [12]:

Coach's intended training load (AU) = intended RPE × intended duration.

Athlete's perceived training load (AU) = RPE × duration.

2.4. Recovery and Stress

Before and after the four weeks of training, RESTQ-sport questionnaires were administered to all speed skaters to define perceptions and sources of recovery and stress [23]. In addition, four weeks prior to the start of the study, a RESTQ-sport was administered to familiarize the speed skaters with the questionnaire. Detailed information about the RESTQ-sport can be found in Kellmann and Kallus [23]. In this study, the Dutch version of the RESTQ-sport was used, which has shown sufficient reliability and validity [24]. The RESTQ-sport was filled out within three days before the start and the end of the four-week monitoring period using the Smartabase athlete management platform (version 6.11.6).

2.5. Inclusion Criteria

Two criteria were used to determine if a speed skater could be included in the analyses. First, the speed skaters needed to complete both RESTQ-sport questionnaires. Secondly, to correct for not filling in the training data, speed skaters had to execute and fill in at least 60% of the intended training load to be included in the analysis. If there was reasonable doubt by one of the researchers in consultation with the coach whether a training log was filled in correctly, all data of that training session were excluded.

2.6. Statistical Analyses

Means and standard deviations of the intended and perceived number of sessions, training load, training load per session, RPE per session, duration, and duration per session were calculated for the four-week monitoring period, compiling all training modalities and for each training modality separately. For these variables, means and standard deviations were also calculated for the different scores of perceived minus intended training variables, which is the difference between coach and athlete. Furthermore, means of the RPE distribution over the four-week period were calculated for intended sessions by the coach and perceived sessions by the speed skaters. To determine differences between intended and perceived training variables, Mann–Whitney U tests were performed for all variables mentioned above. For the different training modalities, percentages of performed training durations were also calculated.

To show differences in the pre- and post-RESTQ-sport scores, means were calculated for the group of speed skaters. To test differences in the pre- and post-RESTQ-sport scores, a Wilcoxon signed-rank test was performed for the aggregated scores of general stress, general recovery, sport-specific stress, sport-specific recovery, total stress, total recovery, and for the recovery–stress balance. For these variables, a difference score between the pre- and post-RESTQ-sport scores was also calculated.

To test if there was a relationship between perceived stress and recovery and differences in training variables intended by the coach and perceived by the speed skaters, correlations between a difference score of the RESTQ-sport measurements and a difference score of training load, RPE per session, and duration per session were calculated. The difference score of the RESTQ-sport was defined as the post-measurement minus the pre-measurement. This was calculated for all variables and for the aggregated variables (i.e., general stress, general recovery, sport-specific stress, sport-specific recovery, and recovery–stress). The difference scores for training characteristics were defined as the perceived value minus the intended value. Before calculating the correlations, the scores were checked for normality. In the case of normality, Pearson correlations were calculated. Otherwise, Spearman correlations were calculated. Descriptive statistics, *t*-tests, and correlations were calculated using SPSS (version 27.0; SPSS, Inc., Chicago, IL, USA). Statistical significance was set at $p < 0.05$. The magnitude of the difference between the measures (effect size) was considered as <0.3 small, 0.3–0.5 moderate, and >0.5 large [25]. The magnitude of correlations are inferred as <0.1 trivial, 0.1–0.3 small, 0.3–0.5 moderate, 0.5–0.7 large, 0.7–0.9 very large, and 0.9–1.0 almost perfect [26].

3. Results

Of the initial 23 speed skaters, 8 did not reach 60% compliance with the intended training load, and 1 speed skater did not complete both RESTQ-sport questionnaires. Consequently, nine speed skaters were excluded for analysis. In total, data from 438 and 378 training sessions were collected from both coaches and speed skaters over four weeks, respectively. Of the remaining 14 speed skaters, 11 executed fewer training sessions than intended (range: 1 to 15 sessions less) and 3 speed skaters executed more sessions than intended (range: 1 to 6 sessions more). Table 1 shows an overview of the intended and perceived training characteristics over the four-week monitoring period.

Table 1. Mean values of intended and perceived training variables (n = 14) along with the difference between training variables (perceived minus intended) over four weeks, mean $\pm$ SD shown. Total refers to summation over the entire 4-week period.

	Intended	Perceived	(Mis)match Perceived-Intended	(Mis)match Range	(Mis)match Effect Size
Total					
Sessions (number)	31 ± 4	27 ± 6	-4 ± 6 *	-15–6	-0.41
Duration (h:min)	$52{:}37 \pm 08{:}41$	$45{:}16 \pm 10{:}09$	$07{:}20 \pm 09{:}09$ *	$-23{:}05$–$07{:}17$	-0.40
Load (AU)	$13{,}686 \pm 2534$	$11{,}609 \pm 2898$	-2076 ± 2615	-6330–3829	-0.31
Average per session					
RPE	4.6 ± 0.4	4.3 ± 0.4	-0.2 ± 0.4	-0.9–0.6	-0.31
Duration (min)	101 ± 5	101 ± 15	0 ± 14	-10–44	-0.17
Load (AU)	435 ± 31	429 ± 57	-7 ± 49	-84–117	-0.01

Note: * $p < 0.05$. AU = Arbitrary Units.

Over the entire four weeks, the intended number of sessions was significantly higher compared to the perceived number of sessions with a moderate effect size ($U = 51.50$, $z = -2.15$, $p = 0.031$, $r = -0.41$), shown in Table 1. The intended training duration was also higher compared to the perceived duration and showed a moderate effect size ($U = 52.00$, $z = -2.11$, $p = 0.035$, $r = -0.40$). Other training variables did not differ over the four weeks, $p > 0.05$.

To provide insight into the differences between and the contributions of the training modalities, Table 2 shows an overview of the intended and perceived training characteristics for the different training modalities. For speed skating, a higher intended versus perceived total training load ($U = 45.00$; $z = -2.44$; $p = 0.014$; $r = -0.46$), RPE per session ($U = 45.00$; $z = -2.45$; $p = 0.014$; $r = -0.46$), and load per session (U = 48.50; $z = -2.28$; $p = 0.021$; $r = -0.43$) were found, showing moderate effect sizes. In addition, for strength training, a higher intended number of sessions ($U = 10.50$; $z = -4.12$; $p < 0.001$; $r = -0.78$), training load ($U = 21.50$; $z = -3.53$; $p < 0.001$; $r = -0.67$), training duration ($U = 7.00$, $z = -4.26$; $p < 0.001$; $r = -0.80$), and duration per session ($U = 52.00$; $z = -2.34$; $p = 0.035$; $r = -0.44$) were found compared to perceived variables along with moderate to large effect sizes. Furthermore, other training modalities showed a moderately higher intended duration per session ($U = 44.50$, $z = -2.46$, $p = 0.014$, $r = -0.46$) and a moderately higher load per session ($U = 53.50$; $z = -2.05$, $p = 0.039$, $r = -0.39$). No differences were found between intended and perceived training characteristics for cycling, $p > 0.05$. Of the total training duration over four weeks, the athletes performed $12.6 \pm 3.4\%$, $52.6 \pm 16.4\%$, $12.6 \pm 3.5\%$, and $22.2 \pm 15.6\%$ of speed skating, cycling, strength training, and other training modalities, respectively.

Table 2. Mean values of intended and perceived training variables for 14 speed skaters along with the difference between training variables (perceived minus intended) for different training modalities over four weeks, mean ± SD shown.

	Intended	Perceived	(Mis)match ΔPerceived–Intended	(Mis)match ΔRange	(Mis)match Effect Size
Speed skating					
Total sessions (number)	5 ± 1	4 ± 1	−1 ± 1	−2–0	−0.28
Total duration (h:min)	06:44	05:41	−01:03 ± 01:20	−03:00–00:20	−0.27
Total load (AU)	2264 ± 550	1733 ± 597	−532 ± 545 *	−1355–340	−0.46
RPE per session	5.7 ± 0.6	5.1 ± 0.5	−0.6 ± 0.7 *	−1.8–0.6	−0.46
Duration per session (min)	79 ± 2	78 ± 5	−1 ± 5	−5–12	−0.15
Load per session (AU)	449 ± 56	396 ± 47	−53 ± 65 *	−147–68	−0.43
Cycling					
Total sessions (number)	14 ± 3	12 ± 4	−2 ± 4	−14–2	−0.21
Total duration (h:min)	27:14	23:34	−03:41 ± 09:39	−29:50–08:02	−0.23
Total load (AU)	6063 ± 1814	5776 ± 2428	−287 ± 2751	−7990–4514	−0.05
RPE per session	3.8 ± 0.6	3.8 ± 0.7	0.0 ± 0.7	−1.8–1.1	−0.02
Duration per session (min)	120 ± 9	120 ± 9	0.0 ± 12	−12–31	−0.14
Load per session (AU)	442 ± 62	475 ± 119	33 ± 128	−211–341	−0.21
Strength training					
Total sessions (number)	7 ± 1	4 ± 2	−3 ± 2 *	−6––1	−0.78
Total duration (h:min)	10:01	05:48	−04:13 ± 02:06 *	−09:15––01:30	−0.80
Total load (AU)	2879 ± 593	1602 ± 704	−1276 ± 530 *	−2160––520	−0.67
RPE per session	4.7 ± 0.7	4.6 ± 0.9	−0.2 ± 0.9	−1.3–1.9	−0.09
Duration per session (min)	89 ± 3	82 ± 9	−6 ± 10 *	−23–12	−0.44
Load per session (AU)	422 ± 57	380 ± 88	−42 ± 83	−146–77	−0.31
Other					
Total sessions (number)	6 ± 2	7 ± 3	1 ± 3	−4–7	−0.24
Total duration (h:min)	08:38	10:14	−01:36 ± 08:00	−06:20–26:35	−0.06
Total load (AU)	2481 ± 658	2499 ± 1519	18 ± 1358	−1880–3350	−0.03
RPE per session	4.8 ± 0.3	4.1 ± 1.4	−0.7 ± 1.4	−5.0–0.7	−0.27
Duration per session (min)	89 ± 6	82 ± 46	−7 ± 48 *	−95–134	−0.46
Load per session (AU)	426 ± 28	348 ± 160	−78 ± 168 *	−470–206	−0.39

Note: * $p < 0.05$. AU = Arbitrary Units.

The RPE distribution over the entire four weeks for all training modalities is shown in Figure 1. Significant differences between the number of intended and perceived sessions on a certain RPE were found on RPE 4, ($U = 46.00$; $z = −2.40$; $p = 0.016$; $r = −0.45$), RPE 6, ($U = 12.00$; $z = −4.01$; $p < 0.001$; $r = −0.76$), and RPE 8 ($U = 53.50$; $z = −2.11$; $p = 0.03$; $r = −0.40$), all with moderate to large effect sizes. In all cases, the intended number of sessions for that RPE was higher compared to the perceived number of sessions (Figure 1).

Figure 2 shows the RESTQ-sport values before and after the four-week monitoring period, and Table 3 shows the differences between the aggregated scores of both RESTQ-sport measurements. Both measurements look comparable upon observation, with lower scores on stress scales compared to higher scores on recovery scales (Figure 2). The only slight variations between pre- and post-RESTQ-sport scores can be viewed for emotional stress, sleep quality, and self-efficacy. This is confirmed by no differences between the pre- and post-RESTQ-sport scores on the aggregated variables (i.e., general stress, general recovery, sport-specific stress, sport-specific recovery, and recovery–stress; range difference: −0.8–0.3; $p > 0.05$).

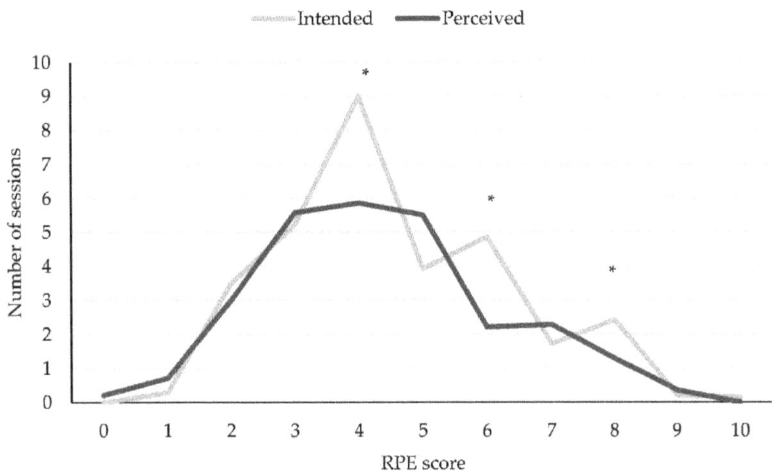

Figure 1. Intended and perceived RPE distribution over all sessions. *: $p < 0.05$.

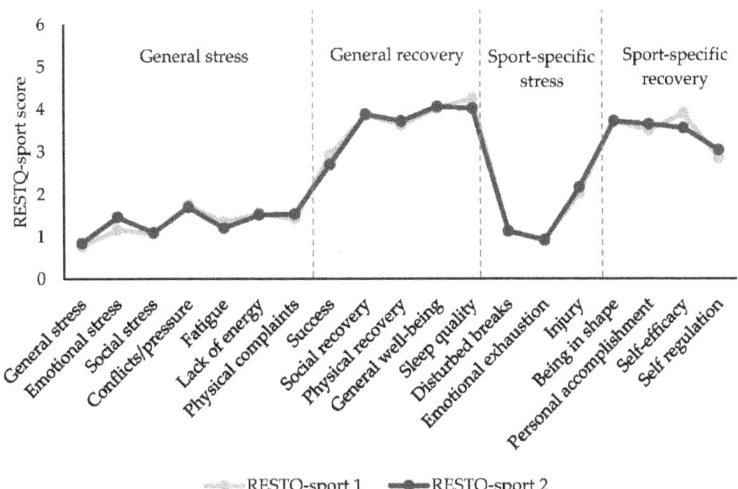

Figure 2. Mean RESTQ-sport scores on the first and second RESTQ-sport measurements. SDs for each subscale range from 0.39 to 1.07. No significant differences were found.

Table 3 shows the correlations between individual changes in RESTQ-sport scores and the difference between intended and perceived training variables. Significant correlations were found between differences in total training load and personal accomplishment, between RPE per session and success, and between duration per session, physical recovery, and self-regulation, $p < 0.05$. All significant correlations were of a large magnitude ($r > 0.56$).

Table 3. Pearson and Spearman correlations between change scores for the RESTQ-sport (post minus pre) and differences between intended and perceived training variables (perceived minus intended) of 14 speed skaters.

	ΔPerceived–Intended			
	Total Sessions	**Total Load**	**RPE Per Session**	**Duration Per Session** [a]
Δ General stress [a]	−0.030	−0.188	0.010	−0.063
Δ Emotional stress	−0.367	−0.475	−0.205	0.056
Δ Social stress	0.049	0.014	−0.210	−0.112
Δ Conflicts/pressure	0.479	0.302	0.090	−0.353
Δ Fatigue	−0.004	−0.200	−0.321	−0.148
Δ Lack of energy	−0.010	−0.060	0.249	−0.259
Δ Physical complaints	−0.247	−0.352	−0.476	0.332
Δ Success	0.309	0.245	0.568 *	−0.118
Δ Social recovery	0.363	0.298	0.117	−0.126
Δ Physical recovery	0.173	0.525	0.389	0.575 *
Δ General well-being [a]	0.233	0.370	0.166	−0.009
Δ Sleep quality	0.038	−0.096	−0.050	−0.186
Δ Disturbed breaks [a]	−0.142	0.052	0.139	0.389
Δ Emotional exhaustion	−0.224	−0.460	−0.275	0.169
Δ Injury	−0.362	−0.411	−0.155	0.197
Δ Being in shape	−0.013	0.048	0.079	0.174
Δ Personal accomplishment	0.391	0.589 *	0.515	0.255
Δ Self-efficacy	0.131	0.313	0.387	0.065
Δ Self-regulation	−0.057	−0.025	−0.043	0.598 *
Δ General stress	−0.053	−0.237	0.024	−0.104
Δ General recovery [a]	0.319	0.354	−0.243	0.106
Δ Sport-specific stress	−0.452	−0.503	0.021	0.357
Δ Sport-specific recovery	0.182	0.380	0.013	0.416
Δ Recovery–stress	0.275	0.437	−0.048	0.077

Note: [a] Spearman correlation; * $p < 0.05$.

4. Discussion

The first aim of this observational study was to examine the (mis)match between the intended training intensity, duration, and load by the coach and the perceived training intensity, duration, and a load of junior speed skaters for training sessions at all intensities and for all training modalities. The second aim was to explore the relationship between the difference between intended and perceived training load and changes in perception of stress and recovery of the speed skaters in a four-week preparatory period.

In summary, the study results showed that the average load and RPE per session were similar for the coach and the speed skaters, though athletes performed, on average, four sessions less in the four-week training period, resulting in lower total training duration, but no difference in the total training load. However, differences between the coach and the individual speed skaters show a very large variation between individual speed skaters with a range of 15 sessions less to 6 sessions more than intended. The training intensity distribution of the sessions reported by the speed skaters was moderate to largely lower than intended by the coach for the RPE scores of 4, 6, and 8. Over the four-week preparatory period, there was no change in perceived stress and recovery on average for the group, but the variation in changes between speed skaters was rather large (SDs ranging from 0.39 to 1.07). The individual change in the RESTQ-sport subscales success, physical recovery, self-regulation, and personal accomplishment showed large positive correlations with the difference between intended and perceived training variables.

This is the first study to show the intended and perceived training intensity, duration, and load of talented junior speed skaters. It was shown that the training load of junior talented speed skaters is lower than what the coach has intended because speed skaters execute fewer sessions. There are no differences in RPE and duration between the coach and the athlete per session including all training modalities. This was also found in a study among elite cyclists [19]; however, a study of young soccer players showed that the athletes trained at a higher RPE than intended [20]. On the other hand, our study shows that the skater's RPE for speed skating sessions is 0.6 lower than what the coach intended. Remarkably, to the best of our knowledge, no other study has shown a lower perceived RPE, on average, than intended by the coach. A reason for the difference may be, for example, because the speed skaters performed their sessions with a lower external load (speed) than intended or because the coach underestimated the physical capacity of the athletes. Additionally, the included skaters were in their late adolescence or early adulthood from an age of 16 to 19 years old. It has been shown that the reliability of the RPE scale improves from childhood to adulthood [9]. Additionally, another study showed that the correlation between the intended RPE of the coach and the perceived RPE of swimmers increases with age and experience [10]. This may explain the large variation of our results, as we included adolescents. Note that it is important that the instructions for how to use the RPE scales are clear to both the coach and speed skaters to align their RPE ratings. In addition, it has been shown that the reliability of RPE can be improved for adolescents when using the OMNI scale [9], and recent findings show that adding facial expressions to the RPE scale is more convenient to use and provides valid and reliable scores for training monitoring [27]. Further research should analyze the reasons why RPE scores from athletes could be lower than those intended by the coach for speed skating sessions and explore the use of facial expressions added to RPE.

Another interesting finding is that the intended training intensity distribution was in line with the training intensity distribution of an Olympic 1500 m medalist [28]. However, the perceived training intensity distribution in our study seems to differ. In 2006 and 2007, the Olympic 1500 m medalist showed a frequency distribution with a peak in the number of sessions at an RPE of four [28]. For the same athlete, a steeper peak was shown in 2008 and 2009 at an RPE of three. Our study shows that the intended training intensity distribution by the coach was similar to the profile of the Olympic medalist in 2006, with a peak at an RPE of four [28]. However, the perceived training intensity distribution is more flattened with a plateau at RPEs of three, four, and five of which the performed sessions at an RPE of four and eight were moderately lower than intended by the coach, and at an RPE of six, were largely lower than intended.

When putting our findings into a three-zone perspective (low, moderate, and high based on RPE < 3, 3–5 and >5, respectively) [14], our study shows that athletes spend less training sessions at moderate and hard intensity than intended by the coaches (see Figure 1). Other studies have shown a higher duration at low intensity, the same or higher duration at moderate intensity, and the same or lower duration at high intensity [14–17]. In our study, training duration was different due to the number of sessions, while other studies showed differences due to a different duration per session. These discrepancies indicate that it is crucial to look at the training duration, intensity, and load as a whole but also at all separate sessions of each individual athlete, coach, and group in order to find a suitable intervention to align the intention and the perception.

A limitation of this study is that only 14 speed skaters out of the initial 23 speed skaters met the inclusion criterium of executing >60% of the intended training load. This resulted in 438 intended sessions by the coach and 378 executed sessions by the skaters. This difference may be due to skaters that did not fill out the training log or did not execute the intended training sessions. In order to ensure as little bias as possible due to not filling out the training log, we have discussed each missing training session with the coach leading to the inclusion criterium of >60%, which ensured that the speed skaters who did not fill out the training log structurally were excluded. Therefore, we are confident that the sessions that were not filled out by the skaters have not been executed. Reasons for not executing these sessions could be a vacation, other obligations, injuries, or illness.

Some studies comparing intended and perceived training load did not report a discrepancy in the number of training sessions [14,15,20] or showed a very small discrepancy [17,19]. If no discrepancy was shown, it was due to the design, which would only include matched sessions. The advantage of including only matching sessions is that a good comparison for each session can be made. However, a clear overview of the total training load that was intended by the coach and the training load that was actually executed by the athletes could not be given, which makes it difficult to make inferences on under- or overtraining. We chose to include all sessions because that gives a realistic overview of the total training load in relation to perceived stress and recovery.

Studies that showed a very small discrepancy included athletes that probably reported their training load comprehensively because it is their profession [19], and because most training sessions were performed under the supervision of the coach [17]. During our study, the junior speed skaters had to do the majority of training sessions on their own initiative such as cycling and other sessions, which leaves more room for absence if they were, e.g., otherwise occupied, or simply not in the mood for training. This is probably the reason why we found differences in the number of sessions between the coach and the athlete. The fact that the coach was not present during most training sessions emphasizes the need for monitoring training load, stress, and recovery in order to give the coach a good overview.

The speed skaters in our study perceived greater success, physical recovery, self-regulation, and personal accomplishment when their training load was approaching or exceeding the intention of the coach. This was shown by a positive correlation between success and the difference between the intended–perceived RPE per session, physical recovery, and self-regulation with perceived–intended duration per session, and personal accomplishment with perceived–intended total training load. However, we did not find significant relationships between differences in perceived–intended training duration, RPE, load, and aggregated scores of total stress and total recovery.

Some studies show a relation between increased training load or a negative life event and success, physical recovery, and self-regulation [29–31]. Remarkably, no other study in individual cyclic sports has shown that personal accomplishment was related to changes in the training load of swimmers and rowers [30–32]. Additionally, no relationship was found between personal accomplishment and changes in the performance of female cyclists [29], nor a change after a negative life event of runners [33]. This suggests that the feeling of personal accomplishment may not be dependent on increases in physical and psychological stress but on the individual matching of training load between coaches and athletes.

Caution should be taken because we included only 14 speed skaters. Therefore, the chance of a false positive correlation is large. Nevertheless, all the scales which show significant correlations are categorized as recovery scales. This points out that the perceived recovery may be a key factor to explain differences between intended and perceived training load, or that differences in intended–perceived training load interfere with perceived recovery.

5. Conclusions

This study found that, during the preparatory phase of 4 weeks, junior speed skaters trained less than was intended by the coach, with no mismatch in RPE and training load calculated for the group. However, we have observed a large variation in the (mis)match, showing a variety between skaters who trained more than intended and skaters who trained less than intended. Additionally, it seems that there are differences between the (mis)match for different training modalities and different training intensities. A (mis)match between the coach and the speed skater is related to changes in recovery scales. That is, when speed skaters approach or exceed the coach's intended RPE, their perception of success seems to improve. Perception of physical recovery and self-regulation improves when they approach or exceed the intended duration per session, and personal accomplishment increases when they approach or exceed the intended training load. This highlights the importance of looking at each individual's deviation from the coach and investigating the reasons for deviation in order to find an appropriate intervention to balance the perception of stress and recovery.

Author Contributions: Conceptualization, R.T.A.O., J.d.J. and I.K.S.; methodology, R.T.A.O., S.v.d.Z. and I.K.S.; validation, S.v.d.Z., J.F.A.G. and I.K.S.; formal analysis, A.C.B., S.v.d.Z. and I.K.S.; investigation, R.T.A.O. and A.C.B.; resources, R.T.A.O.; data curation, A.C.B.; writing—original draft preparation, R.T.A.O. and A.C.B.; writing—review and editing, T.T., I.K.S., S.v.d.Z., J.F.A.G. and J.d.J.; visualization, A.C.B., I.K.S.; supervision, R.T.A.O. and J.d.J.; project administration, T.T. and J.d.J.; funding acquisition, R.T.A.O., J.d.J. and I.K.S. All authors have read and agreed to the published version of the manuscript.

Funding: This research was cofunded by the Stichting Innovatie Aliantie RAAK publiek, RAAK.PUB06.043.

Institutional Review Board Statement: The study was conducted in accordance with the Declaration of Helsinki and approved by the Ethics Committee of the University of Groningen, UMCG, Department of Human Movement Sciences (protocol code ECB/2017.07.21_1R1, 14 January 2022).

Informed Consent Statement: Informed consent was obtained from all subjects involved in the study.

Data Availability Statement: The data presented in this study are available upon reasonable request to the corresponding author.

Acknowledgments: We thank all the speed skaters and their coaches for their support and voluntary participation in this study. In addition, we thank Fleur Smith, Janco Nolles, Rick Nijland, Berry van Holland, and Dick van Dijk for their support with the data collection and analysis.

Conflicts of Interest: The authors declare no conflict of interest.

References

1. Orie, J.; Hofman, N.; de Koning, J.J.; Foster, C. Thirty-Eight Years of Training Distribution in Olympic Speed Skaters. *Int. J. Sports Physiol. Perform.* **2014**, *9*, 93–99. [CrossRef]
2. Stoter, I.; Elferink-Gemser, M.T. High Speed on the Ice: Talent Development in Dutch Long Track Speed Skating. In *Talent Identification and Development in Sport*; Routledge: London, UK, 2020; pp. 158–169.
3. Meeusen, R.; Duclos, M.; Foster, C.; Fry, A.; Gleeson, M.; Nieman, D.; Raglin, J.; Rietjens, G.; Steinacker, J.; Urhausen, A. Prevention, Diagnosis and Treatment of the Overtraining Syndrome: Joint Consensus Statement of the European College of Sport Science (ECSS) and the American College of Sports Medicine (ACSM). *Eur. J. Sport Sci.* **2013**, *13*, 1–24. [CrossRef]
4. Drew, M.K.; Finch, C.F. The Relationship Between Training Load and Injury, Illness and Soreness: A Systematic and Literature Review. *Sports Med.* **2016**, *46*, 861–883. [CrossRef] [PubMed]
5. Matos, N.F.; Winsley, R.J.; Williams, C.A. Prevalence of Nonfunctional Overreaching/Overtraining in Young English Athletes. *Med. Sci. Sports Exerc.* **2011**, *43*, 1287–1294. [CrossRef]
6. American College of Sports Medicine (Acsm). *ACSM's Guidelines for Exercise Testing and Prescription*, 9th ed.; Lippincott Williams Wilkins: Philadelphia, PA, USA, 2013.
7. Borresen, J.; Ian Lambert, M. The Quantification of Training Load, the Training Response and the Effect on Performance. *Sports Med.* **2009**, *39*, 779–795. [CrossRef] [PubMed]
8. Impellizzeri, F.M.; Marcora, S.M.; Coutts, A.J. Internal and External Training Load: 15 Years On. *Int. J. Sports Physiol. Perform.* **2019**, *14*, 270–273. [CrossRef]
9. Groslambert, A.; Mahon, A.D. Perceived Exertion: Influence of Age and Cognitive Development. *Sports Med.* **2006**, *36*, 911–928. [CrossRef]
10. Barroso, R.; Cardoso, R.K.; Carmo, E.C.; Tricoli, V. Perceived Exertion in Coaches and Young Swimmers with Different Training Experience. *Int. J. Sports Physiol. Perform.* **2014**, *9*, 212–216. [CrossRef]
11. Roete, A.J.; Elferink-Gemser, M.T.; Otter, R.T.A.; Stoter, I.K.; Lamberts, R.P. A Systematic Review on Markers of Functional Overreaching in Endurance Athletes. *Int. J. Sports Physiol. Perform.* **2021**, *16*, 1065–1073. [CrossRef]
12. Foster, C.; Hector, L.L.; Welsh, R.; Schrager, M.; Green, M.A.; Snyder, A.C. Effects of Specific versus Cross-Training on Running Performance. *Eur. J. Appl. Physiol.* **1995**, *70*, 367–372. [CrossRef]
13. Foster, C.; Boullosa, D.; McGuigan, M.; Fusco, A.; Cortis, C.; Arney, B.E.; Orton, B.; Dodge, C.; Jaime, S.; Radtke, K.; et al. 25 Years of Session Rating of Perceived Exertion: Historical Perspective and Development. *Int. J. Sports Physiol. Perform.* **2021**, *16*, 612–621. [CrossRef]
14. Foster, C.; Heimann, K.M.; Esten, P.L.; Brice, G.; Porcari, J.P. Differences in Perceptions of Training by Coaches and Athletes. *South Afr. J. Sports Med.* **2001**, *8*, 3–7.
15. Inoue, A.; do Carmo, E.C.; de Souza Terra, B.; Moraes, B.R.; Lattari, E.; Borin, J.P. Comparison of Coach-Athlete Perceptions on Internal and External Training Loads in Trained Cyclists. *Eur. J. Sport Sci.* **2021**, *8*, 1261–1267. [CrossRef]
16. Wallace, L.K.; Slattery, K.M.; Coutts, A.J. The Ecological Validity and Application of the Session-RPE Method for Quantifying Training Loads in Swimming. *J. Strength Cond. Res.* **2009**, *23*, 33–38. [CrossRef] [PubMed]
17. Barnes, K.R. Comparisons of Perceived Training Doses in Champion Collegiate-Level Male and Female Cross-Country Runners and Coaches over the Course of a Competitive Season. *Sports Med. Open* **2017**, *3*, 38. [CrossRef]

18. Borg, G. Perceived Exertion as an Indicator of Somatic Stress. *Scand. J. Rehabil. Med.* **1970**, *2*, 92–98.
19. Voet, J.G.; Lamberts, R.P.; de Koning, J.J.; de Jong, J.; Foster, C.; van Erp, T. Differences in Execution and Perception of Training Sessions as Experienced by (Semi-) Professional Cyclists and Their Coach. *Eur. J. Sport Sci.* **2021**, 1–9. [CrossRef]
20. Brink, M.S.; Frencken, W.G.P.; Jordet, G.; Lemmink, K.A.P.M. Coaches' and Players' Perceptions of Training Dose: Not a Perfect Match. *Int. J. Sports Physiol. Perform.* **2014**, *9*, 497–502. [CrossRef]
21. Kenttä, G.; Hassmén, P. Overtraining and Recovery: A Conceptual Model. *Sports Med.* **1998**, *26*, 1–16. [CrossRef]
22. Stoter, I.K.; Koning, R.H.; Visscher, C.; Elferink-Gemser, M.T. Creating Performance Benchmarks for the Future Elites in Speed Skating. *J. Sports Sci.* **2019**, *37*, 1770–1777. [CrossRef]
23. Kellmann, M.; Kallus, K.W. *Recovery-Stress Questionnaire for Athletes: User Manual*; Human Kinetics: Champaign, IL, USA, 2001.
24. Nederhof, E.; Brink, M.S.; Lemmink, K.A. Reliability and Validity of the Dutch Recovery Stress Questionnaire for Athletes. *Int. J. Sport Psychol.* **2008**, *39*, 301–311.
25. Field, A. *Discovering Statistics Using IBM SPSS Statistics*; Sage: Newcastle upon Tyne, UK, 2013.
26. Hopkins, W.G.; Marshall, S.W.; Batterham, A.M.; Hanin, J. Progressive Statistics for Studies in Sports Medicine and Exercise Science. *Med. Sci. Sports Exerc.* **2009**, *41*, 3–12. [CrossRef] [PubMed]
27. van der Zwaard, S.; Hooft Graafland, F.; van Middelkoop, C.; Lintmeijer, L.L. Validity and Reliability of Facial Rating of Perceived Exertion Scales for Training Load Monitoring. *J. Strength Cond. Res.* **2022**. *accepted*. [CrossRef] [PubMed]
28. Orie, J.; Hofman, N.; Meerhoff, L.A.; Knobbe, A. Training Distribution in 1500-m Speed Skating: A Case Study of an Olympic Gold Medalist. *Int. J. Sports Physiol. Perform.* **2021**, *16*, 149–153. [CrossRef] [PubMed]
29. Otter, R.; Brink, M.; van der Does, H.; Lemmink, K. Monitoring Perceived Stress and Recovery in Relation to Cycling Performance in Female Athletes. *Int. J. Sports Med.* **2015**, *37*, 12–18. [CrossRef] [PubMed]
30. González-Boto, R.; Salguero, A.; Tuero, C.; González-Gallego, J.; Márquez, S. Monitoring the Effects of Training Load Changes on Stress and Recovery in Swimmers. *J. Physiol. Biochem.* **2008**, *64*, 19–26. [CrossRef] [PubMed]
31. Jürimäe, J.; Mäestu, J.; Purge, P.; Jürimäe, T. Changes in Stress and Recovery after Heavy Training in Rowers. *J. Sci. Med. Sport* **2004**, *7*, 335–339. [CrossRef]
32. Nicolas, M.; Vacher, P.; Martinent, G.; Mourot, L. Monitoring Stress and Recovery States: Structural and External Stages of the Short Version of the RESTQ Sport in Elite Swimmers before Championships. *J. Sport Health Sci.* **2019**, *8*, 77–88. [CrossRef] [PubMed]
33. Otter, R.; Brink, M.; Diercks, R.; Lemmink, K. A Negative Life Event Impairs Psychosocial Stress, Recovery and Running Economy of Runners. *Int. J. Sports Med.* **2015**, *37*, 224–229. [CrossRef]

International Journal of
Environmental Research and Public Health

MDPI

Article

Monitoring the Changing Patterns in Perceived Learning Effort, Stress, and Sleep Quality during the Sports Training Period in Elite Collegiate Triathletes: A Preliminary Research

Yi-Hung Liao [1], Chih-Kai Hsu [1,†], Chen-Chan Wei [2,†], Tsung-Chieh Yang [3], Yu-Chi Kuo [1], Li-Chen Lee [4], Li-Ju Lin [1,5,*] and Chung-Yu Chen [6,*]

[1] Department of Exercise and Health Science, National Taipei University of Nursing and Health Sciences, Taipei 112, Taiwan; yihungliao.henry@gmail.com (Y.-H.L.); lout0629@gmail.com (C.-K.H.); yuchi.kuo@gmail.com (Y.-C.K.)
[2] Department of Aquatic Sports, University of Taipei, Taipei 111, Taiwan; tom911072@gmail.com
[3] Dodo Bird Exotic Animal Hospital, Hsinchu 300, Taiwan; centromerevm@gmail.com
[4] Office of Physical Education, Shih-Hsin University, Taipei 116, Taiwan; lilychen@mail.shu.edu.tw
[5] Smart Healthcare Interdisciplinary College, National Taipei University of Nursing and Health Sciences, Taipei 112, Taiwan
[6] Department of Exercise and Health Sciences, University of Taipei, Taipei 111, Taiwan
* Correspondence: liju@ntunhs.edu.tw (L.-J.L.); fish0510@gmail.com (C.-Y.C.)
† These authors contributed equally to this work.

Citation: Liao, Y.-H.; Hsu, C.-K.; Wei, C.-C.; Yang, T.-C.; Kuo, Y.-C.; Lee, L.-C.; Lin, L.-J.; Chen, C.-Y. Monitoring the Changing Patterns in Perceived Learning Effort, Stress, and Sleep Quality during the Sports Training Period in Elite Collegiate Triathletes: A Preliminary Research. *Int. J. Environ. Res. Public Health* **2022**, *19*, 4899. https://doi.org/10.3390/ijerph19084899

Academic Editors: Rafael Oliveira and João Paulo Brito

Received: 15 March 2022
Accepted: 17 April 2022
Published: 18 April 2022

Publisher's Note: MDPI stays neutral with regard to jurisdictional claims in published maps and institutional affiliations.

Abstract: Background: Few studies have examined the mental profiles and academic status of collegiate triathletes during training/competitive periods. We evaluated the changes in sleep quality, physical fatigue, emotional state, and academic stress among collegiate triathletes across training periods. Methods: Thirteen collegiate triathletes (19–26 years old) were recruited in this study. Mood state, sleep quality, degree of daytime sleepiness, subjective fatigue, and academic learning states were measured during the following five training periods: before national competitions for 3 months (3M-Pre Comp), 2 months (2M-Pre Comp), 1 month (1M-Pre Comp), 2 weeks (2wk-Pre Comp), and national competition (Comp) according to their academic/training schedule. Results: The academic stress index in 1M-Pre Comp (Final exam) was significantly higher than that in 3M-Pre Comp in these triathletes. No markedly significant differences were observed in overall mood state, sleep quality, individual degree of sleepiness, and fatigue among these five periods. However, the profiles mood state scale (POMS)-fatigue and -anger were lower in 2wk-Pre Comp than that in 1M-Pre com. The POMS-tension score in Comp was significantly higher than that in 3M-Pre Comp and 2M-Pre Comp. POMS-depression in Comp was lower than that in 1M-Pre Comp. Conclusion: We found that training volume was highest one month before a competition, and the academic stress is greatest during their final term exam period (1M-Pre Comp). After comprehensive assessment through analyzing POMS, PSQI, ESS, and personal fatigue (CIS), we found that the collegiate triathletes exhibited healthy emotional and sleep states (PSQI score < 5) across each training period, and our results suggest that these elite collegiate triathletes had proficient self-discipline, time management, and mental adjustment skills.

Keywords: athletes; fatigue; mood; academic learning states

1. Introduction

A triathlon is a competitive sporting event that combines swimming, cycling, and running. Competition distances vary in different triathlon formats [1]. The biggest difference between triathlons and other sporting events is that triathletes spend more time training than other athletes. Each of the three disciplines requires separate extensive training. Consequently, these athletes may be under greater stress physically, psychologically, and socially [2]. Triathlon training must be tailored to an individual's needs and condition.

Training periodization and a daily schedule are crucial [2]. Training periodization is a means of helping athletes reach their physical and result goals and involves the scheduling of long, moderate, and short training cycles and sessions [3]. Periodization can help triathletes achieve peak levels of regulation and maximum performance during the most important event of the year. Athletes and their coaches should thus pay attention to the type of physical training adaptation required and use the necessary training or skills to achieve a particular adaptive response to physical training [4].

Triathlons are organized worldwide and year-round and comprise both events for elite athletes and leisure competitions. Long duration training that is too intense, combined with insufficient recovery time, can reverse the many positive physiological adjustments associated with training adaptation, resulting in overtraining [5]. Training intensity affects an athlete's body and psychology to certain levels, and appropriate overtraining can induce athletic performance enhancement. However, an excessive training intensity combined with other stress factors may result in overtraining syndrome (OTS) [6]. Researchers widely believe that endurance athletes are at higher risk of OTS due to the prolonged and massive physical and psychological stress they are under due to, for example, training and competing [7]. Clinical diagnosis of OTS is made when athletes exhibit fatigue, mood changes, sleep deprivation, declining athletic performance, and increasing frequency of injury and pain [7]. However, very few studies have investigated the physical and mental stress on triathletes resulting from training and competing.

School sports events are sports competitions primarily for university athletes and include triathlons. From the perspective of specialized sports training, the literature indicates that sufficient research has been conducted into different training period levels, athletic performance, and physical and psychological stress. For example, studies have compared professional elite athletes [1,8,9] and investigated sex differences in athletic performance [10], athletic performance in specific disciplines [8,10], and changes in physiological traits throughout a season [11–13]. Student athletes, especially elite athletes, need to deal with frequent training and high training intensity and volume in their periodic training schedule. However, they must also spend considerable time and energy maintaining their academic performance [14]. Due to the diverse and massive training intensity for triathlons in particular, we believe that the external and internal loadings on student athletes are not smaller than the physical and mental stress experienced by professional athletes in any other competitive sport. According to surveys, collegiate athletes often experience problems with sleep, emotions, and fatigue [15–17]. Although numerous factors can affect the overall performance of triathletes, research into the effect of specific training factors on burnout among triathletes is currently lacking [11,18]. Crucially, the extreme pressure exerted by a heavy learning workload and intense athletic training at this stage can disrupt student athletes' normal physiological and psychological development. Comprehensive assessments of triathletes' mood, sleep, internal and external loadings, fatigue, and athletic performance for different disciplines and training cycles are relatively lacking. Therefore, in the field of sports medicine, more research into the influence of the type of training periodization on students who engage in long and intensive training is necessary.

We hypothesized that the perceived learning effort, mental stress status, and sleep quality would be negative impacted by the increasing training intensity across different periodic training periods in these elite collegiate triathletes. This study evaluated the differences in training conditions, sleep quality, emotional status, and academic pressure among collegiate triathletes with different training cycles. The possible factors affecting these variables were investigated. The evaluation results are used to explore the physical and psychological changes in collegiate triathletes. Multiple questionnaires through periodic survey were used to assess the physical and mental state of student triathletes across different training periods, as well as provide a comprehensive assessment of their academic stress. Therefore, the findings provide coaches and supervisors in sports sciences with reference data when scheduling training for collegiate triathletes and can assist student athletes to avoid injuries due to intense training and overtraining for competitions.

2. Materials and Methods

2.1. Participants and Ethical Statement

The participants in this study were 13 collegiate triathletes (5 females, 8 males) between the ages of 19–26 years. The anthropometric profiles of the participants are shown in Table 1. All participants had fixed formulated annual training schedules under the supervision of their team coach. During this study, the researchers did not change the coaches' training schedule. Participants that met the following conditions were excluded: (1) inability to comply with the training schedule; (2) severe skeletal or muscular injury within the preceding 3 months; (3) mental illness; (4) heart disease, diabetes, or other metabolic diseases. The researchers explained the experimental process to the collegiate triathlon coaches and then announced the recruitment through posters. Before the study officially started, the researchers explained the research process and precautions to all participants. Participants were required to sign informed consent forms. This study was performed according to the last version of the Helsinki Declaration and approved by the Institutional Review Board (IRB) at the University of Taipei (IRB-2018-066).

Table 1. Participants' anthropometric profiles.

Measurement	Male (*n* = 8)	Female (*n* = 5)
Age (year)	20.0 ± 0.3	21.6 ± 0.7
Height (cm)	172.6 ± 1.8	158.4 ± 0.9
Weight (kg)	68.0 ± 2.5	56.7 ± 1.7
Body Mass Index (BMI, kg/m^2)	22.8 ± 0.7	22.7 ± 0.8
Muscle mass (kg)	52.2 ± 1.5	37.9 ± 0.8
Body fat percentage (%)	19.6 ± 0.01	30.4 ± 0.01

2.2. Study Design and Procedure

To minimize any effect exerted by the students' previous training, the coaches avoided scheduling intense exercise and resistance training for 2 days prior to physical performance tests. The questionnaire comprised a sleep quality scale, mood scale, daytime sleepiness scale, and fatigue questionnaire. From 7:00 to 8:30 am on the day of the periodic survey, participants completed four different questionnaires after measuring anthropometric test, including the Profiles Mood State (POMS) scale for mood state evaluation, Pittsburgh Sleep Quality Index (PSQI) scale for sleep quality evaluation, Checklist individual strength (CIS) for subjective fatigue, and Epworth Sleepiness Scale (ESS) scale for daytime sleepiness evaluation. To ensure adequate dietary habit control throughout the training period, we gave the athletes detailed nutritional guidelines. Based on their individual daily energy demands and training period, the participants were instructed to consume the following food groups in these proportions: carbohydrates, 60–65%; protein, 15–20%; and fat, 15–20%, according to the previous study focusing on triathletes [19]. The detailed procedure and timeframe of this study are illustrated in Figure 1.

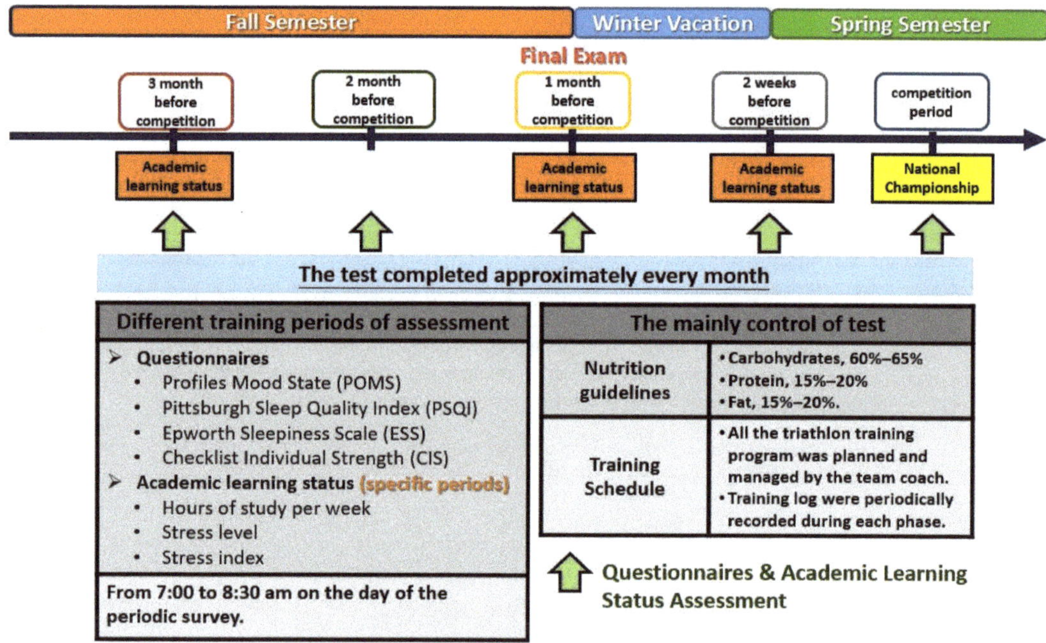

Figure 1. The collegiate triathlete periodic training program detailed procedure and time frame.

2.3. Anthropometric Measurements

The participants were requested to avoid any form of exercise for 2 days before the anthropometric measurements and fast for 10 h before the anthropometric test. All instruments were calibrated according to the manufacturers' directions before the testing. Height was measured using a Dong Sahn Jenix height scale (Seoul, Korea). Weight, body fat percentage, fat-free weight, body mass index, and other anthropometrics were measured using the OMRON HBF-371 body composition monitor (Kyoto, Japan) based on the bioelectrical impedance method. The participants were requested to wear the same light clothing to all measurement sessions to ensure consistency across their pretest and posttest bioelectrical impedance analysis (BIA) measurements. Further, all participants were instructed to prohibit from drinking water and any fluids containing caffeine 2 h prior to the BIA assessment to minimize possible interference and ensure consistency. To ensure the reliability of the BIA measurements, the analysis was conducted with the participants standing in their bare feet on the electrode plate with their hands on the electrode handles.

2.4. Triathletes Periodic Training Program

The triathletes were also asked to maintain their training logs during their periodic training periods, and the training log was the training diary recorded by their team coach to record individual athletes' training type, frequency, intensity, and training feedback. These logs were used to calculate their training intensity for specific sports disciplines. Figure 2 shows the training program during the different training periods. The triathlon training consisted of running, swimming, cycling, and strength/brick training. All participating athletes were performing their regular training under the coach's supervision during the periodic training program. The weekly training time for the four training periods was 824 min (3 months before competition, 3M-Pre Comp), 1010 min (2 months before competition, 2M-Pre Comp), 1175 min (1 month before competition, 1M-Pre Comp), and 786 min (2 weeks before competition, 2wk-Pre Comp). During the data collection period, it can be seen that the majority of training hours were spent on swimming and cycling, and the training duration for these two sports disciplines was about 1:1, followed by running and then weight training. While the brick training took up the least amount of total training, it can still be seen that the training time proportion for brick training increased 2 weeks before the national competition events.

2.5. Perceived Learning Effort and Study Time

Participants were asked to rate their level of perceived learning effort on a Likert scale of 1 (very easy) to 10 (extremely stressful) for their regular subjective studying effort and academic preparation during their specific training periods. In addition, participants were asked to provide the average number of study hours per week during the specific training period for the perceived learning work effort assessment. Banister et al. calculated the training impulse (TRIMP) by multiplying the exercise duration by the exercise intensity, which mainly reflects the overall exercise volume during exercise [20]. Our investigation applied this concept into consideration and designed an academic stress index calculated by multiplying perceived learning effort by studying duration to represent overall academic stress/volume across different training periods. In brief, the triathlete academic stress index during each training period was calculated using the product of the subjective perceived learning effort level during their academic activities and the average number of hours of study per week for that training cycle.

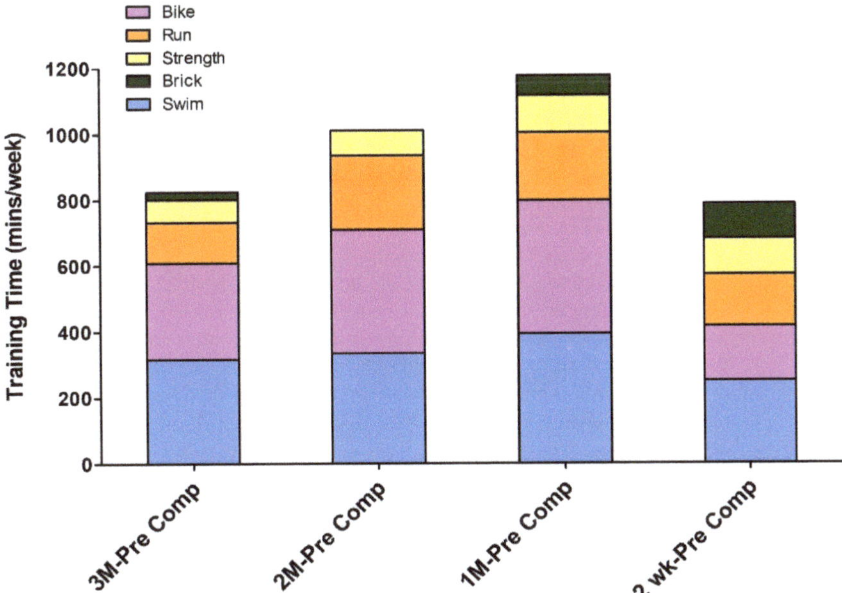

Figure 2. Training schedule for the different training periods. 3M-Pre Comp: 3 months before competition, 2M-Pre Comp: 2 months before competition, 1M-Pre Comp: 1 month before competition, 2wk-Pre Comp: 2 weeks before competition.

2.6. Profiles Mood State Questionnaire (POMS)

The POMS was used to measure the emotional state of the athletes under specific training and competition conditions. The POMS questionnaire was developed by McNair et al. and comprises 37 questions covering seven elements [21]. The elements vigor (7) and self-esteem (5) represent positive moods, whereas the elements confusion (7), fatigue (6), anger (5), tension (4), and depression (3) represent negative moods. Each question was graded on a 5-point Likert scale from 0 (not at all) to 4 (extremely), and the score for each item was directly added to the total. When answering each question, the participants responded to the question "How have you felt for the past 2 weeks (including today)?".

2.7. Pittsburgh Sleep Quality Index (PSQI)

The Chinese version of the PSQI scale, as translated by Tsai et al., was used in this study [22]. The PSQI scale has nine questions covering seven dimensions: sleep quality, sleep latency, sleep time, habitual sleep efficiency, sleep disorders, sleep medication use, and daytime dysfunction. Each dimension is worth 0–3 points; the maximum number of points is 21. A higher PSQI score indicates poorer sleep quality. Typical scores are less than or equal to 5, signifying favorable sleep quality. The participants completed the scale according to their sleep during the previous month. The Cronbach's α for the scale is 0.8332. The Chinese PSQI validity was verified using the PSQI scale tests. The questionnaire's accuracy, sensitivity, and specificity were found to be 88.5%, 89.6%, and 86.5%, respectively ($\kappa = 0.75$; $p < 0.001$), demonstrating that the Chinese PSQI scale has satisfactory reliability and validity.

2.8. Epworth Sleepiness Scale

The Epworth Sleepiness Scale (ESS) is a questionnaire scale used to assess the degree of daytime sleepiness of participants [23]. The ESS is a subjective evaluating scale that allows participants to self-assess the probability of dozing in eight different situations (e.g., sitting and reading, watching television, etc.) on a scale ranging from 0 (no probability of dozing) to 3 (high probability of dozing). The scores were mainly calculated by summing the scores of the eight questions and the total score represented their daytime sleepiness. A score in the range of 0–9 was considered normal, while a score in the range of 10–24 indicated excessive drowsiness.

2.9. Checklist Individual Strength (CIS)

The Checklist Individual Strength (CIS) questionnaire was used to assess the subjective fatigue level of triathletes at each training period. The CIS questionnaire consists of 20 items, and each question is scored on a 7-point Likert scale. The total score for the test was obtained by summing the item scores. These 20 items were spread over four subscales: fatigue severity, which measures (i) subjective fatigue (8 items); (ii) concentration, which measures attention problems (5 items); (iii) motivation, which measures decreased motivation (4 items); and (iv) physical activity, which measures decreased activity (3 items). In the CIS questionnaire, some items were reverse-scored for more precise assessment. CIS exhibits reliable internal consistency ($\alpha = 0.84$–0.95) [24].

2.10. Statistical Analysis

SPSS 16.0 (Chicago, IL, USA) and GraphPad Prism 5.0 (La Jolla, CA, USA) were used to analyze and graph the data, respectively. SPSS 16.0 was employed to perform the Shapiro–Wilk normality test to analyze the normality of the variables of interest. One-way analysis of variance and repeated measures were used to compare changes in academic status, emotional status, subjective fatigue, daytime sleepiness, and sleep quality during the overall training and competition periods. All data were expressed as the mean $\pm$ standard error of the mean, and the level of significance in all comparisons was set as 0.05 ($p < 0.05$).

3. Results

3.1. Effects of Different Training Periods on Academic Stress Status of Triathletes

Figure 3 presented the athlete's academic learning status results during different training periods. Figure 3A displays the subjective weekly study hours. The weekly study hours were significantly higher in 1M-Pre Comp (Final exam) compared with 3M-Pre Comp. However, 2 wk-Pre Comp presented significantly lower weekly study hours compared with 3M-Pre Comp and 1M-Pre Comp (Final exam) ($p < 0.05$; $\eta^2 = 0.226$). There were no differences among the three phases for subjective study stress in these athletes (Figure 3B).

The academic stress index in 1M-Pre Comp (Final exam) was significantly higher than that in 3M-Pre Comp in this population (Figure 3C) ($p < 0.05$; $\eta^2 = 0.109$).

3.2. Effects of Different Training Periods on Mood States in Triathletes

POMS questionnaire was used to assess the mood state and is shown in Figure 4. There were no significant differences in overall mood states (Figure 4A), anger (Figure 4B), self-esteem (Figure 4C), and confusion (Figure 4D) among each period in these athletes. However, the fatigue (Figure 4E) and anger (Figure 4F) scores were significantly lower in the 2 wk-Pre comp compared with the 1M-Pre comp ($p < 0.05$; fatigue: $\eta^2 = 0.022$; anger: $\eta^2 = 0.065$). The tension scores were significantly higher in Comp than in 3M-Pre Comp (3 month before competition) and 2M-Pre Comp (Figure 4G) ($p < 0.05$; $\eta^2 = 0.148$). The level of depression during competition period (Comp) was significantly lower than in 1M-Pre Comp (Figure 4H) ($p < 0.05$; $\eta^2 = 0.152$).

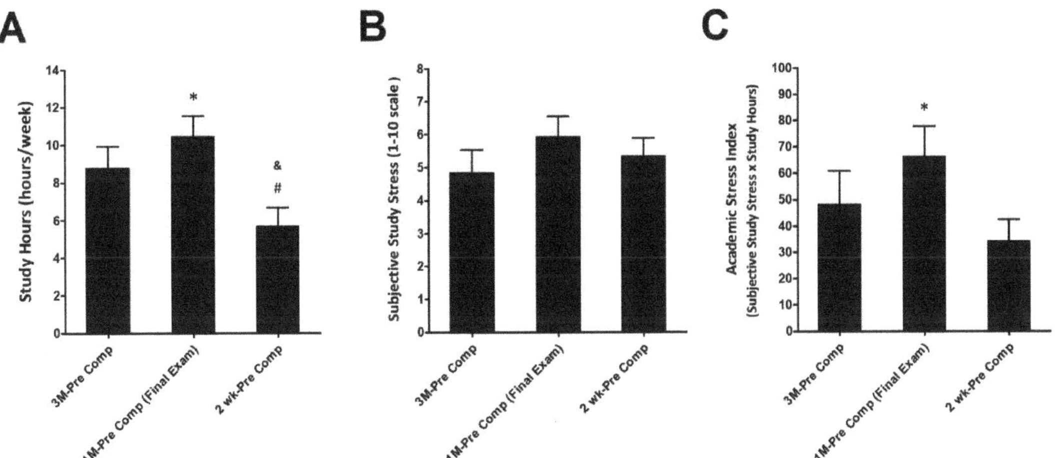

Figure 3. Subjective academic learning effort and stress status. (**A**) Study hours, (**B**) subjective study stress, and (**C**) academic stress index (assessed by multiplying study stress and study hours) were measured and collected at three training period. Data are represented as Mean ± S.E.M. * denotes significant difference between 3M-Pre Comp and 1M-Pre Comp (Final exam) ($p < 0.05$). # denotes significant difference between 3M-Pre Comp and 2wk-Pre Comp ($p < 0.05$). & denotes significant difference between 1M-Pre Comp (Final exam) and 2wk-Pre Comp ($p < 0.05$). See the legend of Figure 2 for abbreviations.

3.3. Effects of Different Training Periods on Degree of Fatigue and Sleepiness in Triathletes

The individual degrees of fatigue and sleepiness assessed by Checklist Individual Strength (CIS) questionnaire and Epworth Sleepiness scale (ESS), respectively, are shown in Figure 5. No significant difference in individual fatigue was observed among these five different training phases (Figure 5A). Additionally, we did not observe any significant differences in degree of sleepiness among the five different training periods in this population (Figure 5B).

3.4. Effects of Different Training Periods on Sleep Quality in Triathletes

Using the PSQI questionnaire to evaluate sleep quality is shown in Figure 6. The overall sleep quality scores in these triathletes for the five different training periods (3M-Pre Comp, 2M-Pre Comp, 1M-Pre Comp, 2wk-Pre Comp, Comp) were not significantly different (Figure 6A) ($p > 0.05$). In addition, there were no significant differences among periods in subjective sleep quality (Figure 6B), sleep latency (Figure 6C), sleep time (Figure 6D), sleep disorders (Figure 6E), and daytime dysfunction (Figure 6F) ($p > 0.05$).

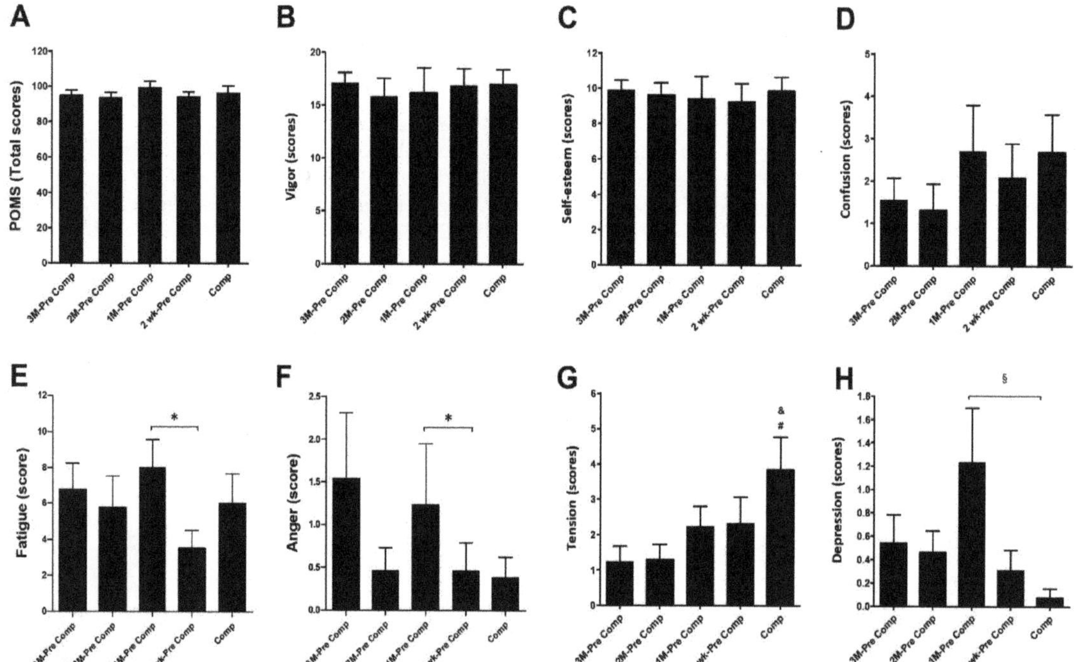

Figure 4. Profile of Mood state. (**A**) Total scores of POMS, (**B**) Vigor of POMS, (**C**) Self-esteem of POMS, (**D**) Confusion of POMS, (**E**) Fatigue of POMS, (**F**) Anger of POMS, (**G**) Tension of POMS, and (**H**) Depression of POMS were measured in five training periods. Data are represented as Mean ± S.E.M. * denotes significant difference between 1M-Pre Comp and 2wk-Pre Comp ($p < 0.05$). & denotes significant difference between 3M-Pre Comp and Comp ($p < 0.05$). # denotes significant difference between 2M-Pre Comp and Comp ($p < 0.05$). § denotes significant difference between 1M-Pre Comp and Comp ($p < 0.05$). See the legend of Figure 2 for abbreviations.

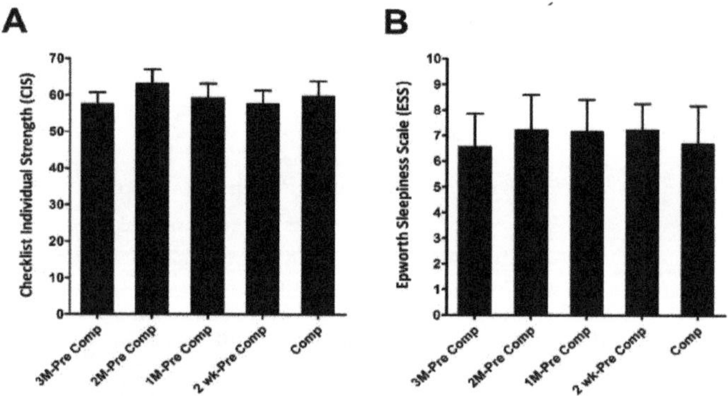

Figure 5. Individual degree of fatigue and sleepiness. Checklist individual strength (**A**) and Epworth sleepiness scales (**B**) were collected at five training periods. Data are represented as Mean ± S.E.M. See the legend of Figure 2 for abbreviations.

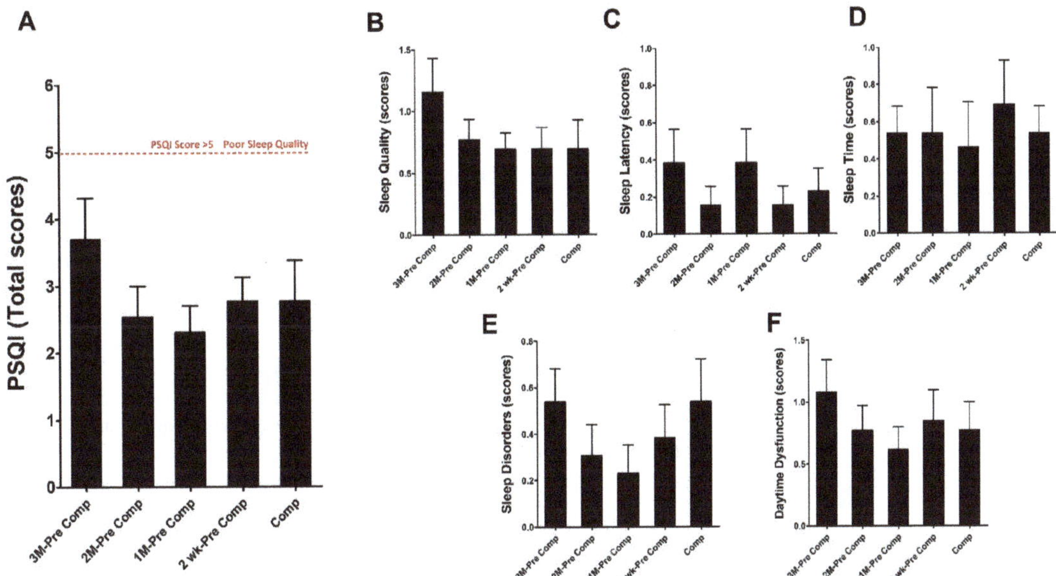

Figure 6. Pittsburgh Sleep Quality Index (PSQI). (**A**) Total PSQI, (**B**) PSQI Sleep quality, (**C**) PSQI Sleep latency, (**D**) PSQI Sleep time, (**E**) PSQI Sleep disorders, and (**F**) PSQI Daytime dysfunction scores were determined in five training periods. Data were represented as Mean ± S.E.M. See the legend of Figure 2 for abbreviations.

4. Discussion

The major finding of this study was that the total number of hours that the collegiate triathletes spent in training sessions—which comprised running, swimming, cycling, weight training, and brick training—gradually increased initially. The total training amount then gradually decreased after reaching a peak 1 month before a competition. Furthermore, more training hours were spent in swimming and cycling than in the other training types. The academic stress index, which was assessed by multiplying study stress and study hours, exhibited that these athletes experienced both greater academic and sports training intensity one-month before national competition events (final term exam period). The ESS and personal fatigue scale results revealed no significant differences between training months during the investigation period. The PSQI results showed no significant difference in the total score for each month (each month's score was less than 5 points), nor any significant difference for the PSQI scale subindexes, namely, subjective sleep quality, sleep latency, habitual sleep efficiency, total sleep hours, sleep disorders, sleep medication use, and daytime dysfunction. The POMS scale results revealed that the total emotional score did not differ significantly between the monitored months. However, the individual POMS scores indicated that the tension index was higher during competition months than other training months. The depression index was highest 1 month before a competition and significantly higher than that in the competition period. The fatigue index was highest 1 month before a competition and, at this time point, it was significantly higher than that 2 weeks before the competition period. No significant differences were discovered in the other POMS assessment indexes—vigor, self-esteem, and anger. Despite the extensive amount of training that the participants underwent during the group training 1 month before a competition, the personal fatigue index scores were not significantly higher during this time.

Triathlon training plans involve running, swimming, cycling, and weight training. In keeping with the basic concepts of the three triathlon stages, the training for each discipline was incorporated into the following annual competition periods: off season, preseason, and competition season [4]. Researchers have demonstrated that in nonelite triathletes, training for 8–10 h total per week (5–6 h cycling and 3–4 h running per week) resulted in the lowest likelihood of injury [18]. Furthermore, during the off season, triathletes focus on mental and physical rest and emphasize resistance training to improve their athletic performance and prevent injury through improving muscle strength [2]. To improve aerobic endurance, training plans must be designed to strengthen the respiratory, cardiovascular, and musculoskeletal system functions, such as through physical strength training to improve athletic performance in specific disciplines [4]. The average periodic training hours for the collegiate triathletes participating in this study were 4–6.5 h of swimming, 3–7 h of cycling, and 2–4 h of running, which are largely equivalent to the hours recommended. The training schedule also included weight training each month, the average duration of which was 1.5 h. One study demonstrated that the triathletes' training period and daily schedule are also critical [2]. As competition season approaches, training periods begin with high-volume, low-intensity training and progresses to low-volume, high-intensity training [4]. The present study discovered that the number of brick training hours increased with the approach of national level competitions. The number of hours for other types of training also increased but peaked 1 month before the competition before decreasing. This demonstrated that the training schedules for collegiate triathletes conform to periodic training patterns. However, in terms of athletic intensity, the physical activity measurements performed in this study did not indicate any significant differences between periods. As the rate of perceived exertion (RPE) and session RPE (sRPE) were not measured, this study was unable to fully understand training intensity. Future studies should use RPE and sRPE to analyze differences in intensity between training periods in greater detail.

Training intensity monitoring is necessary for understanding an individual's response to training stimuli and assessing the levels of fatigue during different periods. The present study discovered that training intensity and training tension were significantly greater during intensive training periods and were significantly positively correlated with the total fatigue score. This indicated that subjective fatigue surveys are a sensitive tool and can be used to perceive changes in training intensity [5]. The Checklist Individual Strength subjective fatigue scale was used in this study and revealed a lack of significant differences in personal fatigue between each training month during the study period. Although the athletes had periodic training schedules, the study results demonstrated that changes to the training volume each month did not significantly influence the fatigue felt by the athletes. According to our results, periodic training schedules do not create significant changes in collegiate triathletes' subjective fatigue. Another possible explanation is that triathletes undergo massive amounts of training in each training period; these athletes are thus fatigued for long periods of time and do not receive sufficient relief from fatigue.

Previous evidence has revealed that the athletes' sleep time and quality dropped after their training intensity increased (+30%), demonstrating an inverse correlation between training intensity and sleep time and quality [25]. Athletes are often unable to obtain sufficient sleep due to training, competing, competition scheduling, travel, stress, and other training intensities and times or due to overtraining. Many other factors may also lead to insufficient sleep in overtrained athletes [25,26]. For example, academic pressure or work burdens and even psychological pressure can negatively affect athletes' energy [17]. The PSQI assessments of triathletes' sleep during this study revealed that the total scores were lower than 4 points and the assessments for each period were not significantly different. The subindexes—sleep quality, sleep latency, habitual sleep efficiency, total sleep time, sleep disorders, sleep medication use, and daytime dysfunction—did not exhibit significant differences, demonstrating that collegiate triathletes commonly experienced good and long sleep. Furthermore, we discovered that the athletes trained the most but had the lowest sleep quality index scores at 1 month before a competition, indicating that the collegiate triathletes had higher sleep quality when they were training more. However, this is inconsistent with studies reporting that an increase in training intensity leads to lower sleep quality and total amount of sleep [25,26]. We noted that 1 month before a competition was during the winter holiday break, when the athletes' academic pressure was relatively low; consequently, the athletes were better able to fully recover (including sleep) from high-intensity training during this period. The present results also revealed that the competition season and training schedule for student athletes may need to be planned with comprehensive consideration of their school semester schedules to ensure they have time to rest and recover while balancing academic and training demands.

Interestingly, during the months with the greatest training volume, only the POMS scale depression and fatigue scores increased. The personal fatigue scale scores did not increase. This was possible due to the collegiate athletes' peak training period coinciding with the winter holiday break—when the athletes could obtain sufficient rest. Therefore, the high POMS fatigue score may have been the result of the athlete's mental state, which could have been affected by psychological factors such as the athlete's own self-demands and their coach's demands. Athlete's fatigue may thus originate from psychological pressure while training rather than from increased physical activity. Training hours were consistent with the durations suggested in the literature, and the few differences may have been because the literature suggestions were for nonelite athletes. The study participants were collegiate elite athletes, and no severe sports injuries or mental stress were discovered during this study. This indicates that the training volume and stress from the coaches did not affect the athletes. This also substantiates the argument of Slivka et al. that by limiting the stress on athletes from sources unrelated to training, athletes can withstand very high training volume and intensity without developing OTS [27].

POMS questionnaire and PSQI scale scores (measuring, e.g., sleep disorders and sleep latency) are considered early indicators of training maladaptation [26]. Studies have found that loss of sleep may have negative effects on subjective well-being measurements, including measurements of fatigue, mood, and confusion. Furthermore, competition stress and anxiety can negatively affect sleep quality and duration, and sleep loss can damage mood and increase stress and anxiety [26]. One study reported that athletes with lower self-assessed sleep quality have higher levels of emotional confusion, leading to a higher risk of failure. Each point increase in confusion level was discovered to lead to 19.7% lower sleep quality [27]. Hausswirth et al. assessed the POMS score and VO_2max of 40 triathletes and continually monitored the participants using Actiwatch. Nine triathletes were diagnosed with functional overtraining with reduced functionality and high perceived fatigue [28]. A significant time–group interaction was also discovered between sleep duration, sleep efficiency, and inactive periods [28]. Romyn et al. also argued that greater tension and anxiety lead to lower sleep quality ratings [29]. Based on the studies above [26], the POMS and PSQI questionnaires can be used to evaluate early indicators of training maladaptation. However, observing only the changes in the total scores of these two questionnaires may neglect the detailed mental changes in these college triathletes. Likewise, our results revealed no significant differences in the total POMS and PSQI scores of athletes across training periods (Figures 4 and 6). Importantly, we found that these athletes showed a significant increase in POMS stress levels during the competition period, yet there were no significant differences in their sleep quality PSQI scores among periods. Our obtained result seems to exhibit certain diversities from the above studies [27–29], and we speculate that these collegiate triathletes might have better attitude when controlling their life pattern, thereby exhibiting greater capacity to cope with mental stress of training or competition by obtaining better sleep quality.

Most studies have used the ESS, Competition Stressor Scale, Checklist for Sleep Hygiene, and PSQI scale to assess sleep and stress in athletes [15,16,30]. Mah et al. reported on inappropriate sleep times among collegiate athletes due to insufficient sleep hours and low sleep quality, and these athletes were often discovered to exhibit visible daytime sleepiness [15]. Reasons given in the literature for daytime sleepiness and poor sleep quality include high-intensity training, extra working time, overuse of electronic devices, skipping breakfast, sports injuries, overtraining, anxiety prior to sleep, and other problems that cause anxiety, pain, and sleep disorders [15,30]. However, it has to be noted that the ESS results in the present investigation revealed no significant differences between training months. The individual assessments revealed that although most of the athlete participants did not experience daytime sleepiness, some athletes had an average ESS score of 16–18 points, indicating that they experienced higher rates of daytime sleepiness. One study suggested that the training schedule prepared by sports team coaches can be personalized by being flexibly adjusted to the athlete's needs and that practical guidance and suggestions given to athletes should lessen their perceived fatigue [17]. Based on the present findings, we suggest that collegiate triathlon coaches conduct individual interviews with athletes with higher sleepiness scores, change their training start time, or provide appropriate sleep-related health education to help these athletes reduce their daytime sleepiness.

Changes in athletes' mental state or mood while training for super-endurance competitions can lead to changes in their performance [31]. Triathletes are considered a group at risk of developing a series of adverse health results due to high training intensity and long training times [18]. Additionally, mood is directly correlated with training amount and intensity [31]. Comotto et al. used sRPE and the POMS scale to monitor a group of teenage elite triathletes (age: 18 ± 1 years) who were given the same external training intensity during training camp. The POMS scores revealed that the athletes' fatigue increased by 45% and vigor decreased by 24%, and the energy index for all training camp athletes decreased [32]. The POMS scores in a 6-month study on 32 triathletes (18 athletes, 15 members in the sedentary control group, all aged 24–61 years) discovered that vigor score decreased and anger and fatigue scores were high during peak training months [31]. Unlike prior studies that included participants with a wide age range, this study focused on collegiate triathletes and used the POMS scale to assess changes in their mood during 4 months of periodic training (this period included general physical preparations and national level competition stages). The results from the present study revealed that the fatigue scores among the collegiate triathletes were significantly higher when the training was most extensive; however, their anger scores were not significantly higher and their vigor scores were not significantly lower. Comparing the present findings with those of a previous study [31], we surmise that emotional stress was not significantly higher when training intensity was higher because of the similarities in age, periodic training schedule, and competition preparations among the collegiate triathletes. Furthermore, school sports teams have peer support, professional coach supervision, and team goals; these attributes could have helped to alleviate the stress caused by intensive training.

This study discovered that although the amount that collegiate triathletes trained differed depending on the particular training period, the differences did not significantly affect the athletes' sleep quality or emotional response. This further substantiates the earlier statement that precompetition preparations that coincide with the winter holiday break allowed athletes to have sufficient rest despite an increase in training volume and greater depression, fatigue, and tension during the competition periods. These results show that collegiate triathletes have healthy emotion and sleep management and also that the reduction in academic pressure during the precompetition period may explain the athletes' favorable sleep quality and mood. Although this study discovered no significant differences between training periods in the overall averages for sleep quality, daytime sleepiness, and emotional stress, some athletes did have significantly higher scores than other athletes in these portions of the questionnaire. These athletes may have had insufficient rest or recovery due to personal factors. Therefore, coaches should provide individualized guidance and recommendations to athletes whose scores indicate poor sleep quality or high stress or refer them to appropriate mental, nutritional, or sleep consultants.

4.1. Study Limitation

One of the primary limitations of the present study is that the participants were limited to national, college-level elite triathletes; therefore, all participants were trained and competed on the same college triathlon team. In this regard, we made every effort in the study design to eliminate potential confounding factors, including standardized dietary instruction by the team coach and the conducted supervised triathlon-specific training program. Despite our efforts to control the study design quality, we still could not completely rule out the possibility of a Type II error due to the relatively small sample size. However, the primary population in this study was collegiate Division I elite triathletes (all of them were ranked Top 10 in the nation in their competition category), which is a specific category and a very limited number of participants could be recruited. On the other hand, although both male and female athletes were included in this study; we did not further analyze the differences between males and females due to the small sample size. Therefore, future studies investigating gender differences in this issue are warranted.

4.2. Perspectives in Practical Application

The results of this study provide crucial information for triathlon training. (1) Questionnaire surveys (e.g., POMS and PSQI scales) have reliability and can reflect athletes' physical and psychological status. Coaches should ask their athletes to regularly complete short personal surveys. (2) Triathlons are competitions for individuals, and different athletes have different physical and psychological statuses during training. Coaches must provide individualized instructions or adjustments based on the problems experienced by the particular athlete. (3) Coaches should consider student athletes' academic stress state and training quality to individually adjust the training intensity and periodic program to achieve greater sports performance.

5. Conclusions

The emotional state, level of fatigue, and sleep quality of 13 collegiate triathletes going through periodic training were collected in this study and assessed over 3 months before the national competitive event. This study discovered that training volume was highest 1 month before a competition. The POMS overall emotional assessment scores indicated no significant differences between different training periods. However, anxiety scores rose during the competition season, while fatigue and depression scores rose during the peak training period. The personal fatigue scores indicated no significant changes in the fatigue level during the investigation period or any significantly greater fatigue during the peak training period. Changes in the ESS total score were also nonsignificant. The PSQI total score and individual scores were not significantly different, indicating favorable sleep quality among the athletes. A comprehensive view of the assessments used in this study—training volume, POMS, PSQI, ESS, and personal fatigue—indicate that the collegiate athletes exhibited healthy emotional and sleep states (PSQI score < 5) during each training period. The athletes had proficient self-discipline, time management, and mental adjustment skills. Administering a subjective questionnaire might help the coaches to closely consider each athlete's individual demands. When selecting evaluation scales, a simply used questionnaire should be considered to minimize the inconvenience to athletes of completing forms after stressful studying and training intensity. Coaches should also appropriately intervene when the athletes exhibit negative patterns to help them better cope with psychological or physical stress to prevent overtraining.

Author Contributions: Conceptualization, Y.-H.L., L.-J.L. and C.-Y.C.; methodology, Y.-H.L., L.-J.L. and C.-Y.C.; formal analysis, C.-K.H. and C.-C.W.; investigation, Y.-H.L., L.-J.L., T.-C.Y., L.-C.L. and C.-Y.C.; resources, Y.-C.K., L.-C.L., C.-K.H. and C.-C.W.; data curation, Y.-H.L., L.-J.L. and C.-Y.C.; writing—original draft preparation, Y.-H.L., L.-J.L. and C.-Y.C.; writing—review and editing, Y.-H.L. and C.-Y.C.; supervision, L.-J.L.; project administration, Y.-H.L. and C.-Y.C.; funding acquisition, Y.-H.L., L.-J.L. and C.-Y.C. All authors have read and agreed to the published version of the manuscript.

Funding: This work was partially supported by the Ministry of Science and Technology, Taiwan (Grant number 110-2410-H-845-021-MY2 for C.Y.C, 109-2628-H-227-002-MY3 for Y.H.L).

Institutional Review Board Statement: The study was conducted according to the guidelines of the last version of the Declaration of Helsinki and approved by the University of Taipei IRB (IRB-2018-066, date of approval 26 November 2018).

Informed Consent Statement: Informed consent was obtained from all subjects involved in the study.

Data Availability Statement: The data presented in this study are available upon request from the corresponding author.

Acknowledgments: The authors thank the athletes for their participation in this study. We also sincerely appreciate the University of Taipei and National Taipei University of Nursing and Health Sciences for providing the necessary resources and administrative support throughout the study.

Conflicts of Interest: The authors declare no conflict of interest.

References

1. Dolan, S.H.; Houston, M.; Martin, S.B. Survey results of the training, nutrition, and mental preparation of triathletes: Practical implications of findings. *J. Sports Sci.* **2011**, *29*, 1019–1028. [CrossRef] [PubMed]
2. Strock, G.A.; Cottrell, E.R.; Lohman, J.M. Triathlon. *Phys. Med. Rehabil. Clin. N. Am.* **2006**, *17*, 553–564. [CrossRef] [PubMed]
3. Tønnessen, E.; Svendsen, I.S.; Rønnestad, B.R.; Hisdal, J.; Haugen, T.A.; Seiler, S. The annual training periodization of 8 world champions in orienteering. *Int. J. Sports Physiol. Perform.* **2015**, *10*, 29–38. [CrossRef] [PubMed]
4. Harvey Wallmann, R.; James, B.S. An introduction to periodization training for the triathlete. *J. Strength Cond. Res.* **2001**, *23*, 55–64. [CrossRef]
5. Elloumi, M.; Makni, E.; Moalla, W.; Bouaziz, T.; Tabka, Z.; Lac, G.; Chamari, K. Monitoring training load and fatigue in rugby sevens players. *Asian J. Sports Med.* **2012**, *3*, 175–184. [CrossRef]
6. Ten Haaf, T.; Van Staveren, S.; Oudenhoven, E.; Piacentini, M.F.; Meeusen, R.; Roelands, B.; Koenderman, L.; Daanen, H.; Foster, C.; De Koning, J.J. Prediction of functional overreaching from subjective fatigue and readiness to train after only 3 days of cycling. *Int. J. Sports Physiol. Perform.* **2017**, *12* (Suppl. 2), S287–S294. [CrossRef]
7. Carfagno, D.G.; Hendrix, J.C., III. Overtraining syndrome in the athlete: Current clinical practice. *Curr. Sports Med. Rep.* **2014**, *13*, 45–51. [CrossRef]
8. Houston, M.; Dolan, S.; Martin, S. The impact of physical, nutritional, and mental preparation on triathlon performance. *J. Sports Med. Phys. Fit.* **2011**, *51*, 583–594.
9. Olmedilla, A.; Torres-Luque, G.; García-Mas, A.; Rubio, V.J.; Ducoing, E.; Ortega, E. Psychological profiling of triathlon and road cycling athletes. *Front. Psychol.* **2018**, *9*, 825. [CrossRef]
10. Lepers, R. Analysis of Hawaii ironman performances in elite triathletes from 1981 to 2007. *Med. Sci. Sports Exerc.* **2008**, *40*, 1828–1834. [CrossRef]
11. Coutts, A.J.; Wallace, L.K.; Slattery, K.M. Monitoring changes in performance, physiology, biochemistry, and psychology during overreaching and recovery in triathletes. *Int. J. Sports Med.* **2007**, *28*, 125–134. [CrossRef] [PubMed]
12. Maimoun, L.; Galy, O.; Manetta, J.; Coste, O.; Peruchon, E.; Micallef, J.P.; Mariano-Goulart, D.; Couret, I.; Sultan, C.; Rossi, M. Competitive season of triathlon does not alter bone metabolism and bone mineral status in male triathletes. *Int. J. Sports Med.* **2004**, *25*, 230–234. [PubMed]
13. Galy, O.; Manetta, J.; Coste, O.; Maimoun, L.; Chamari, K.; Hue, O. Maximal oxygen uptake and power of lower limbs during a competitive season in triathletes. *Scand. J. Med. Sci. Sports* **2003**, *13*, 185–193. [CrossRef] [PubMed]
14. Pritchard, M. Comparing sources of stress in college student athletes and non-athletes. *Athl. Insight* **2005**, *5*, 1–8.
15. Mah, C.D.; Kezirian, E.J.; Marcello, B.M.; Dement, W.C. Poor sleep quality and insufficient sleep of a collegiate student-athlete population. *Sleep Health* **2018**, *4*, 251–257. [CrossRef]
16. Monma, T.; Ando, A.; Asanuma, T.; Yoshitake, Y.; Yoshida, G.; Miyazawa, T.; Ebine, N.; Takeda, S.; Omi, N.; Satoh, M.; et al. Sleep disorder risk factors among student athletes. *Sleep Med.* **2018**, *44*, 76–81. [CrossRef]
17. Kennedy, M.D.; Tamminen, K.A.; Holt, N.L. Factors that influence fatigue status in Canadian university swimmers. *J. Sports Sci.* **2013**, *31*, 554–564. [CrossRef]
18. Main, L.C.; Landers, G.J.; Grove, J.R.; Dawson, B.; Goodman, C. Training patterns and negative health outcomes in triathlon: Longitudinal observations across a full competitive season. *J. Sports Med. Phys. Fit.* **2010**, *50*, 475–485.
19. Nogueira, J.A.; Da Costa, T.H. Nutrient intake and eating habits of triathletes on a Brazilian diet. *Int. J. Sport. Nutr. Exerc. Metab.* **2004**, *14*, 684–697. [CrossRef]
20. Banister, E.W.; Macdougall, J.D.; Wenger, H.A. *Modeling Elite Athletic Performance: Physiological Testing of the High-Performance Athlete*; Human Kinetics: Champaign, IL, USA, 1991.
21. McNair, D.M.; Lorr, M.; Droppleman, L.F. *Manual for the Profile of Mood States*; Educational and Industrial Testing Services: San Diego, CA, USA, 1971.
22. Tsai, P.S.; Wang, S.-Y.; Wang, M.-Y.; Su, C.-T.; Yang, T.-T.; Huang, C.-J.; Fang, S.-C. Psychometric evaluation of the Chinese version of the Pittsburgh Sleep Quality Index (CPSQI) in primary insomnia and control subjects. *Qual. Life Res.* **2005**, *14*, 1943–1952. [CrossRef]

23. Johns, M.W. A new method for measuring daytime sleepiness: The Epworth sleepiness scale. *Sleep* **1991**, *14*, 540–545. [CrossRef] [PubMed]
24. Worm-Smeitink, M.; Gielissen, M.; Bloot, L.; van Laarhoven, H.; van Engelen, B.; van Riel, P.; Bleijenberg, G.; Nikolaus, S.; Knoop, H. The assessment of fatigue: Psychometric qualities and norms for the Checklist individual strength. *J. Psychosom. Res.* **2017**, *98*, 40–46. [CrossRef] [PubMed]
25. Lastella, M.; Vincent, G.; Duffield, R.; Roach, G.D.; Halson, S.; Heales, L.J.; Sargent, C. Can sleep be used as an indicator of overreaching and overtraining in athletes? *Front. Physiol.* **2018**, *9*, 436. [CrossRef] [PubMed]
26. Watson, A.M. Sleep and Athletic Performance. *Curr. Sports Med. Rep.* **2017**, *16*, 413–418. [CrossRef]
27. Slivka, D.R.; Hailes, W.S.; Cuddy, J.S.; Ruby, B.C. Effects of 21 days of intensified training on markers of overtraining. *J. Strength Cond. Res.* **2010**, *24*, 2604–2612. [CrossRef]
28. Hausswirth, C.; Louis, J.; Aubry, A.; Bonnet, G.; Duffield, R.; Meuret, Y.L.E. Evidence of disturbed sleep and increased illness in overreached endurance athletes. *Med. Sci. Sports Exerc.* **2014**, *46*, 1036–1045. [CrossRef]
29. Romyn, G.; Robey, E.; Dimmock, J.A.; Halson, S.L.; Peeling, P. Sleep, anxiety and electronic device use by athletes in the training and competition environments. *Eur. J. Sport Sci.* **2016**, *16*, 301–308. [CrossRef]
30. Hoshikawa, M.; Uchida, S.; Hirano, Y. A subjective assessment of the prevalence and factors associated with poor sleep quality amongst elite japanese athletes. *Sports Med. Open* **2018**, *4*, 10. [CrossRef]
31. Jason, A.J. Time course of performance changes and fatigue markers during training for the ironman triathlon. In *Department of Human Kinetics and Ergonomics 2010*; Rhodes University: Grahamstown, South Africa, 2009.
32. Comotto, S.; Bottoni, A.; Moci, E.; Piacentini, M.F. Analysis of session-RPE and profile of mood states during a triathlon training camp. *J. Sports Med. Phys. Fit.* **2015**, *55*, 361–367.

International Journal of
Environmental Research and Public Health

Article

Single- and Multi-Joint Maximum Weight Lifting Relationship to Free-Fat Mass in Different Exercises for Upper- and Lower-Limbs in Well-Trained Male Young Adults

Danilo A. Massini [1], Anderson G. Macedo [1,2], Tiago A. F. Almeida [2], Mário C. Espada [3,4,*], Fernando J. Santos [3,4,5], Eliane A. Castro [1,2,6], Daniel C. P. Ferreira [1], Cassiano M. Neiva [1,2,7] and Dalton M. Pessôa Filho [1,2]

1 Graduate Programme in Human Development and Technology, São Paulo State University (UNESP), Rio Claro 13506-900, Brazil; dmassini@hotmail.com (D.A.M.); andersongmacedo@yahoo.com.br (A.G.M.); elianeaparecidacastro@gmail.com (E.A.C.); daniel.crisfe@unesp.br (D.C.P.F.); merussi.neiva@unesp.br (C.M.N.); dalton.pessoa-filho@unesp.br (D.M.P.F.)
2 Department of Physical Education, São Paulo State University (UNESP), Bauru 17033-360, Brazil; tiagofalmeida.w@gmail.com
3 School of Education, Polytechnic Institute of Setúbal, 2914-504 Setubal, Portugal; fernando.santos@ese.ips.pt
4 Life Quality Research Centre, 2040-413 Rio Maior, Portugal
5 Faculty of Human Kinetics, University of Lisbon, 1499-002 Lisbon, Portugal
6 LFE Research Group, Department of Health and Human Performance, Faculty of Physical Activity and Sport Sciences, Universidad Politécnica de Madrid (UPM), 28040 Madrid, Spain
7 MEFE—Metabolism and Exercise Physiology Laboratory, Faculty of Science, São Paulo State University (UNESP), Bauru 17033-360, Brazil
* Correspondence: mario.espada@ese.ips.pt; Tel.: +351-265-710-800

check for **updates**

Citation: Massini, D.A.; Macedo, A.G.; Almeida, T.A.F.; Espada, M.C.; Santos, F.J.; Castro, E.A.; Ferreira, D.C.P.; Neiva, C.M.; Pessôa Filho, D.M. Single- and Multi-Joint Maximum Weight Lifting Relationship to Free-Fat Mass in Different Exercises for Upper- and Lower-Limbs in Well-Trained Male Young Adults. *Int. J. Environ. Res. Public Health* **2022**, *19*, 4020. https://doi.org/10.3390/ijerph19074020

Academic Editors: Rafael Oliveira and João Paulo Brito

Received: 19 February 2022
Accepted: 24 March 2022
Published: 28 March 2022

Publisher's Note: MDPI stays neutral with regard to jurisdictional claims in published maps and institutional affiliations.

Abstract: This study aimed to analyze whether the relationship between regional and whole-body fat-free mass (FFM) and strength is related to FFM distribution and area according to limb involvement. Thirty well-trained male young adults underwent one-repetition maximum test (1RM) to assess the strength in arm curl (AC), bench press (BP), seated row (SR), leg press 45° (LP45), knee extension (KE), and leg curl (LC). Dual-energy X-ray absorptiometry was used to evaluate FFM. The values for 1RM in AC, BP, and R correlated to FFM in upper limb (R^2 = 0.69, 0.84 and 0.75), without an effect of appendicular mass index (API) or area. For 1RM in KE, the correlation with FFM in lower limb increased with thigh area (R^2 = 0.56), whereas 1RM in LC and LP45 correlation to whole-body FFM increased with API (R^2 = 0.64 and 0.49). The upper limb's FFM may be reliable for indexing the arms and upper trunk strengths, whereas the relationships between FFM and strength in lower limb improve as muscle mass and thigh area increases between subjects.

Keywords: muscle strength; resistance exercise; body composition; early adulthood

1. Introduction

Resistance exercise promotes muscular fitness (i.e., an increase in muscle strength and work economy, and improvement in power and speed during daily living or sporting tasks), which is undoubtedly accompanied by physiological and morphological muscle adaptations [1–3]. Nonetheless, muscle adaptation to resistance training requires that variables are planned (choice of exercise, order of exercise, load, volume, rest, frequency, and repetition velocity) to match a specific goal [2,4,5]. Indeed, when dealing with advanced practitioners (i.e., many years of training), further improvements in strength and muscle hypertrophy require the adequate management of training variables (e.g., load, repetition, sets, rest, and motor task) during a single session or throughout planning [1].

The loading in resistance training is operationally defined as the percentage of one-repetition maximum weight lifted (%1RM) in a single- or multi-joint exercise [4,5]. There are existing protocols for the measurement of the 1RM value [6]; however, these procedures are unreasonable when considering the training routine and planning for advanced practitioners, which include higher %1RM, high training volume (multiple sets, and a high number of repetitions), and high frequency to encompass a variety of single- and multi-joint exercises [1,5]. Alternatively, the monitoring of 1RM in terms of body composition and anthropometry is supported by the assumption that muscle strength increases in association with the modifications of fat-free mass (FFM) and, therefore, also influencing lift performance.

It was previously shown that segmental body area (arm circumference, arm muscle cross-sectional area, and thigh circumference) also makes a significant contribution to strength in highly resistance-trained athletes [7,8]. Moreover, the fewer joints and muscle groups involved in a weight lifting session, the greater the predictive accuracy from variables of body dimensions. However, the power of this relationship is controversial among studies [9,10]. Hortobágyi et al. [9] concluded that individual differences in muscular strength are poorly related to various measures of body size and segmental body dimensions, since correlations between strength vs. body mass, FFM, thigh and arm volume, cross-sectional area, and skinfolds ranged from −0.52 to 0.56 for trained and non-trained subject groups. Conversely, Hetzler et al. [10] evidenced improvements in the estimate of 1RM bench press using the repetitions to failure test with the addition of the arm circumference and arm length.

In the earliest studies reporting the relationship between 1RM values and anthropometric information, the coefficients widely ranged, but were not above 0.9 [8,11–15]. Therefore, when collectively analyzed, most of these previous studies have related sectional and muscle areas, circumference, and body mass to 1RM performance in multi-joint exercises (i.e., bench press and squat), resulting in predictive equations without the same robustness of the estimate as the models considering the submaximal level of muscle strength (i.e., repetition to failure based on a given weight, percentage of body mass, or fixed number of lifts) [16]. However, an improvement in correlation coefficient has been reported when FFM is considered as an independent variable to be related with the strength for exercises engaging single joints and small muscle groups [8,14] regardless of the level of training (i.e., moderate or advanced) of the participants [7,9,11,17].

Information is surprisingly lacking regarding the power of regional composition to monitor the 1RM value, despite findings indicating the influence of physical performance, FFM, and muscle fiber hypertrophy on the ability to lift heavier weight [18,19]. Indeed, if regional body tissue adaptations are considered to be meaningful information, combined with whole-body changes, and with practical (re)considerations for training control and planning across sexes and ages [20], it would be interesting to analyze how the regional composition information may be useful to evaluate the variations in 1RM in exercises regarding muscle mass participation in resistance exercises.

Thus, the objective of this study was to analyze whether regional and whole-body FFM, which are expected to correlate with 1RM in upper- and lower-limb exercises, follow a specific tendency concerning the limb engaged in exercise. In addition, we wondered whether regional FFM influences the change in 1RM values according to the differences in anthropometric and other composition variables between participants. We hypothesized that regional FFM correlation with 1RM values follows a specific trend regarding the limb engaged in the lift movement, therefore presenting a stronger coefficient compared to anthropometric and whole-body FFM variables. In others words, confirmation that strength and FFM are more strongly related at the body region level will demonstrate that muscle force and mass are both parameters of limb enhancement or a decreased ability in lifting exercises. This would support training and rehabilitation plans regarding body region requirements for strength improvements.

2. Materials and Methods

2.1. Participants

Thirty well-trained male adult volunteers (23.7 ± 5.8 years, 178.7 ± 5.3 cm in height, 78.7 ± 11.3 kg in body weight, and 17.0 ± 5.4% in body fat), with resistance training experience of at least two years and no injury episode during the last six months, provided their written informed consent to participate in this study. Only male young adults participated to avoid the interference of maturation, sex, and aging process on muscle strength, fat-free tissue mass, and bone mineral content among subjects [17,21]. This research was approved by the Local Ethics Committee of the University (CAEE: 19824719.3.0000.5398).

2.2. Body Composition

The dual-energy X-ray absorptiometry (DXA) method (Hologic® model, QDR Discovery Wi®, Beldford, MA, USA) was used to obtain the regional and whole-body composition. The software (Hologic APEX®, Beldford, MA, USA) provided values of FFM (fat-free mass and bone mineral content, in grams) for upper and lower limbs (UL-FFM and LL-FFM), and the submaximal whole-body FFM (WB-FFM, discarding values for the head). Other regional and whole-body composition variables were fat mass (FM), area, and appendicular fat-free mass index (API). The equipment was calibrated following the manufacturer's recommendations by a laboratory technician with experience in these procedures. According to Nana et al. [22], the standardized conditions for DXA scanning are: (i) participants be presented fasted, rested (no exercise), and with no fluid ingestion for at least three hours before the analysis, and (ii) should arrive wearing light clothing, without shoes or carrying any metallic object or body-worn accessories. During the DXA scanning, the participants remained lying in the supine position on the table until the end of the scan, with feet kept together (~15 cm apart) and arms arranged along the side of the trunk (in a midprone position with ~3 cm between the palms and trunk). The same technician adjusted the anatomical points following the manufacturer recommendations. The participants underwent DXA scanning during the first visit.

2.3. Strength Measurements

Tests of 1RM were performed on the following exercises: (1) arm curl (AC), (2) horizontal bench press (BP), (3) seated row (SR), (4) knee extension (KE), (5) leg curl (LC), and (6) leg press 45° (LP45). All tests were performed after a non-specific warm-up of 15 min (static stretching, cycling, or running at exercise intensity ≤60% age-predicted maximal heart rate (i.e., HRmax = 220 − age, with age in years). The protocol of the 1RM test followed the recommendations of Mayhew et al. [23]: (1) a specific warm-up preceded the first attempt of the test and was performed with light weights to avoid concentric failure, and up to 8–10 non-maximal repetitions; (2) initial test weight was chosen based on the average rates for the strength of upper- and lower-limbs, according to age, sex, and body mass [6]; and (3) participants performed at least three attempts of one repetition each, with three minutes of rest between each attempt. The weight was increased or decreased from the initial weight by 1.1 to 4.5 kg based on the difficulty of the first lift. The weight that could not be lifted twice (i.e., self-reported inability, or failure in attempt, to perform the second lift) represented the 1RM reference [6,7]. The load value was reported in kilograms (kg). The participants were instructed to perform the movements with the proper technique, following recommendations from Baechle and Earle [24]. Moreover, two visits, separated by 24 h, were scheduled for the completion of all 1RM testing, following the order of small to large muscle groups, intercalating upper- and lower-limb exercises. Thus, AC, KE, and BP were tested in the first visit, and LC, SR, and LP45 in the second visit. Participants were instructed to avoid high-intensity resistance training 48 h before the testing, and to present themselves rested, fasten, and well-hydrated two hours prior to testing.

2.4. Statistical Analysis

The data are reported as mean ± standard deviation, confidence interval ($CI_{95\%}$), and standard error of measurement (SEM). Normality was checked for the muscle strength variables by the Shapiro–Wilk test. The Pearson coefficient (r) was used to test the linear relationship (2-tailed) between maximum observed strength and body composition variables. The stepwise method was used to model the linear relationship between values of 1RM (as the dependent factor) and regional and whole-body composition variables (as independent factors). The input data for muscle strength in UL exercises considered regional and whole-body composition variables, except those for LL, and vice versa when the procedures were applied to analyze the relationship between LL strength exercises and body composition. To ensure that the correlations were not inflated for the differences in muscle area and musculature distribution, the analysis was controlled to segment area (i.e., arm or thigh, according to the body region involved in the exercise) and API (independently of the body region involved in the exercise). The Pearson coefficient was interpreted as <0.2 (trivial), 0.20–0.49 (small), 0.5–0.8 (medium), and >0.8 (strong) [24]. Scatterplots was used to analyze the explained variance (R^2 and R^2_{adj}) and standard error of the estimate (SEE) of the FFM-predicted 1RM distribution to the observed 1RM distribution across subjects, considering both coefficients as <0.04 (trivial), 0.04–0.24 (small), 0.25–0.63 (medium), and >0.64 (strong) [25]. All statistical procedures were performed in SPSS 26 (Statistical Package for Social Sciences, IBM, Armonk, NY, USA), with a significance level of $p \leq 0.05$.

The sample power for the associations between the observed and predicted 1RM values were determined for each exercise, and the mean value was considered for analysis of the sample size ($n = 30$). Input parameters were: (a) the corresponding value of "*r*" from the coefficient for explained variance (R^2) given in scatterplots; (b) $Z\alpha = 1.96$ for a security index of $\alpha = 0.05$, following Díaz and Fernandéz [26]:

$$Z_{1-\beta} = \sqrt{n-3}\,\frac{1}{2}ln\left(\frac{1+r}{1-r}\right) - Z_{1-\frac{\alpha}{2}} \tag{1}$$

To avoid anon-realistic statistical power by using information from the actual sample, the cross-validation process was performed using the predicted residual error sum of squares (*PRESS*) method [27,28]. From the *PRESS* statistic, a modified form of R^2 adjusted (R^2_p) and standard error of the estimate (SEE_p) were recalculated, $R^2_p = 1 - (PRESS/SS_{Total})$ and $SEE_p = (PRESS/n)^{1/2}$, in which *PRESS* is the sum of the squares of eliminated residuals:

$$PRESS = \sum_{i=1}^{n}(y_i - \hat{y}_{i,-i})^2 \tag{2}$$

3. Results

Table 1 presents regional and whole-body composition characteristics, anthropometric area, and 1RM values of the participants.

Table 1. Values of regional and whole-body composition, and muscle strength.

		Mean ± SD	**CI$_{95\%}$**	**SEM**
Index	API (kg/m^2)	9.67 ± 1.02	9.30–10.06	0.19
FFM	WB (g)	59,596.8 ± 6831.6	57,045.9–62,147.8	1247.3
	UL (g)	8439.0 ± 1387.8	7920.7–8957.2	253.4
	LL (g)	22,316.1 ± 3064.9	21,171.7–23,460.6	559.6
Areas	Arm (cm^2)	480.1 ± 45.7	463.0–497.1	8.3
	Thigh (cm^2)	837.9 ± 95.5	802.2–873.6	17.4
Exercises 1RM	AC (kg)	44.8 ± 9.3	41.3–48.3	1.7
	BP (kg)	82.5 ± 21.2	74.6–90.7	3.9
	SR (kg)	96.5 ± 23.2	87.9–105.2	4.2
	KE (kg)	133.4 ± 29.1	122.5–144.2	5.3
	LC (kg)	90.8 ± 20.7	83.1–98.6	3.8
	LP45 (kg)	323.6 ± 61.5	300.6–346.6	11.2

API: appendicular fat-free mass index; FFM: fat-free mass; WB: whole-body; UL upper limbs; LL: lower limbs; AC: arm curl; BP: horizontal bench press; SR: seated row; KE: knee extension; LC: leg curl; LP45: leg press 45°; 1RM: one-repetition maximum; SD: standard deviation; CI$_{95\%}$: confidence interval; SEM: standard error of measurement.

The correlation coefficients between the regional and whole-body composition variables with 1RM values are shown in Table 2. All correlation coefficients for UL- and LL-FFM were observed to be at a higher level than those for API, WB-FFM, and arm and thigh areas, with the exceptions of KE, LC, and LP45, for which the correlations with WB-FFM and LL-FFM were quite similar.

Table 2. Coefficients for Pearson's correlation analysis between 1RM values and regional and whole-body composition variables.

Exercises	**Body Composition**					
	API	**WB-FFM**	**UL-FFM**	**LL-FFM**	**Arm**	**Thigh**
AC	0.67 ** [medium]	0.71 ** [medium]	0.82 ** [strong]	na	0.60 ** [medium]	na
BP	0.83 ** [strong]	0.86 ** [strong]	0.91 ** [strong]	na	0.73 ** [medium]	na
SR	0.73 ** [medium]	0.83 ** [strong]	0.86 ** [strong]	na	0.76 ** [medium]	na
KE	0.56 ** [medium]	0.71 ** [medium]	na	0.74 ** [medium]	na	0.65 ** [medium]
LC	0.58 ** [medium]	0.79 ** [medium]	na	0.77 ** [medium]	na	0.72 ** [medium]
LP45	0.60 ** [medium]	0.68 ** [medium]	na	0.63 ** [medium]	na	0.50 ** [small]

API: appendicular fat-free mass index; WB: whole-body; FFM: fat-free mass; UL: upper limbs; LL: lower limbs; AC: arm curl; BP: horizontal bench press; SR: seated row; KE: knee extension; LC: leg curl; LP45: leg press 45°. ** $p < 0.001$, na: not analyzed.

Figure 1 depicts the scatterplots between values for the 1RM tests. For AC, BP, and SR, the explained variances from UL-FFM (Figure 1A–C) were higher when controlled by API (R^2 = 0.69, 0.84, and 0.75, respectively, (strong), $p < 0.01$). A similar result was observed for the KE variance explained by LL-FFM (Figure 1D), which increased when controlled by the thigh area (R^2 = 0.54 (medium), $p < 0.01$), and for the LC and LP45 variances explained by WB-FFM (Figure 1E and F), which also increased when controlled by API (R^2 = 0.62 (strong) and 0.46 (medium), respectively, $p < 0.01$).

The *PRESS* analysis is presented in Table 3. The stability of the correlations by shrinkage analysis from R^2_{adj} to R^2_p was ensured for all observed correlations between RE and

FFM variables, since the values for R^2_p were at a ≤ 0.1 ratio from the previous R^2_{adj} values. Cross-validation was therefore acceptable from R^2_p for regression analysis in all resistance exercises. Calculated unbiased estimates of SEE_p reduced when compared to those SEE shown in Figure 1 for all resistance exercises.

Table 3. Cross-validation values from *PRESS* analysis.

Exercise	Model	Cross-Validation			
	R^2_{adj}	R^2_p	Shrinkage	SEE_p (kg)	SEE_{Dif} (%)
AC	0.66	0.63	0.03	5.54	+2.21
BP	0.83	0.82	0.01	8.92	+2.41
SR	0.73	0.70	0.03	12.44	+2.89
KE	0.53	0.43	0.01	21.56	+7.85
LC	0.61	0.55	0.06	13.71	+5.79
LP45	0.44	0.40	0.04	46.93	+1.89

AC: arm curl; BP: horizontal bench press; SR: seated row; KE: knee extension; LC: leg curl; LP45: leg press 45°; SEE: standard error of estimate. SEE_{Dif}: difference between SEE_p and SEE.

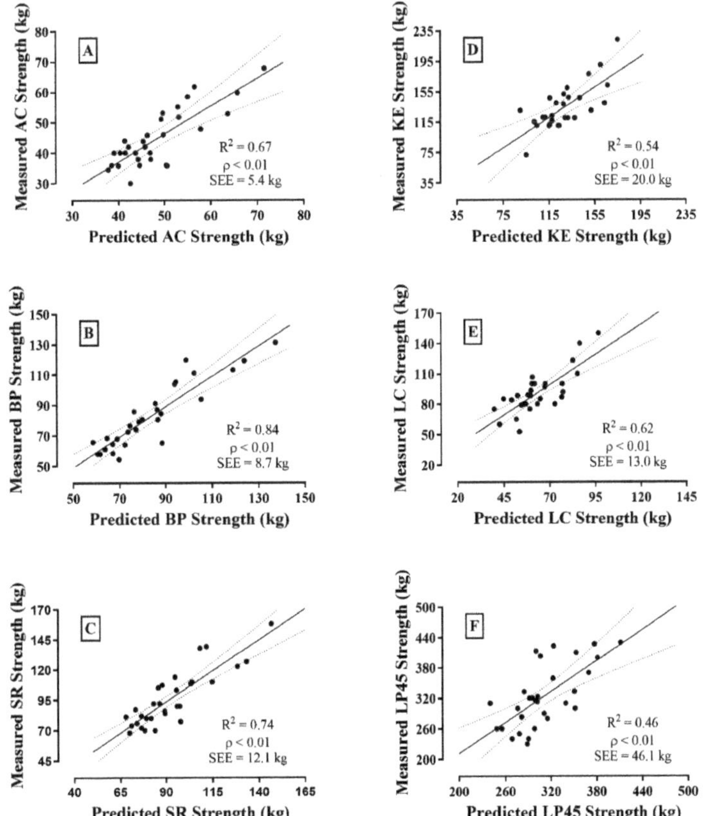

Figure 1. Scatterplots between observed and predicted values for the 1RM tests in bench press (BP) (**A**); arm curl (AC) (**B**); seated row (SR) (**C**); knee extension (KE) (**D**); leg curl (LC) (**E**); and leg press 45° (LP45) (**F**).

4. Discussion

The aim of this study was to analyze whether regional and whole-body FFM follows a specific tendency concerning the limb engaged in exercise. In addition, we wondered whether regional FFM influences the change in 1RM values according to the differences in anthropometric and other composition variables between participants. The findings from the present study showed that both UL- and LL-FFM are powerful indexes that are related to 1RM measurements for single- or multi-joint resistance exercises engaging upper- and lower-limb actions. Therefore, our findings are aligned with the assumption that resistance training can improve muscle strength, weight lifting capacity, and fat-free body mass [2,18]. However, information on the propensity of regional body composition to analyze muscle strength variance in different weight lifting exercises is still lacking in the literature. Thus, the current study evidenced that 1RM correlations with FFM in upper and lower limbs in exercises involving single- and multi-joint actions increased according to the content of FFM, regardless of the peripheral FFM distribution and thigh area between subjects when considering resistance exercises involving upper and lower limbs (respectively).

In this sense, the way that FFM variables related with 1RM values for UL single- or multi-joint exercises evidenced a higher power for regional than whole-body FFM, regardless of the arm sectional area between subjects. Moreover, the LL-FFM is a relevant variable for 1RM values when considering LL lifting weight capacity. However, the LL-FFM variable did not achieve a higher power than the whole-body FFM for the correlation with all resistance exercises. The FFM peripheral distribution (i.e., API) accounted for the increase in the correlation coefficients for LC and LP45; therefore, the results suggest that the greater the engagement of muscle mass for the execution of the exercise, the less the regional influence of FFM seems to be.

Undoubtedly, monitoring 1RM values based on regional FFM is an alternative way to control the muscle strength variation [8,29,30]. Moreover, a successful maximum lifted weight during a standard 1RM test protocol presumes: (i) movement expertise and engagement, (ii) soreness and injury possibilities, and (iii) changes in the weight lifted with the difference in mechanical demand of similar exercises. These are the greatest constraints for the testing protocol frequency and application to every exercise planned for training [29–31]. Therefore, the power of the interactions between maximum weight lifting capacity with body composition parameters (i.e., body mass, fat-free body mass, regional body area and volume, girth, and width) would provide confident references for 1RM measurements, controlling muscle strength improvements, and organizing or revising the overload during the training in accordance with the previous target weight and exercise volume [7,9,10,20,32–34].

However, the literature has shown conflicting results for assessing 1RM using anthropometric and body composition variables, mainly when it is carried out with subjects with differences in muscle strength. On the one hand, results showing that among trained subjects, anthropometric variables (arm circumference and length) improved the reliability (R^2 changed from 0.87 to 0.90) of 1RM estimation in the bench press [10]. Additionally, the predictive power (multiple regression coefficient, R^2) of the anthropometric dimension variables for 1RM estimates ranged from 0.52 to 0.87 for trained subjects [16,30]. Body composition and anthropometry have been related to variations in muscle strength among untrained subjects, but evidence of associations with 1RM were small to medium (Pearson's coefficient ranging from 0.42 to 0.67), mainly for LL and UL multi-joint resistance exercises [9,13,14,16].

Furthermore, 75.7% of the strength assessed in the bench press by trained men can be explained by the variations in the cross-sectional area of the arm, BMI, and fat percentage, with a standard error of 12.1 kg in the prediction [11]. In addition, the strength in the bench press exercise, in populations of both sexes and varying strength levels, showed a high correlation with the variable lean mass (0.77), and moderate correlations with height (0.59), body weight (0.56), arm circumference (0.66), and chest circumference (0.60), although only the lean mass and submaximal load for 10RM estimated the bench press strength with 97.6% explanatory capacity [8].

Thus, the statement that highly trained athletes exhibit closer relationships between anthropometric dimensions and weight lifted, and probably, the fewer joints and muscle groups involved in a lift, the greater the predictive accuracy of maximum performance by structural proportions [8], remain theories about the association between training development and the responses in the body's dimensions. The results from the current study agree with this statement. Furthermore, we extend this assertion to exercises involving UL, considering that the association was independent of arm area size, but increased with FFM distribution in the upper limb. Moreover, for LL exercises, the control of 1RM values should consider changes in whole-body FFM and its peripheral distribution between subjects.

The lack of research relating regional body composition to 1RM for single-joint resistance exercises, contrast to those analyzing whole-body composition, anthropometry, and sub-maximal lifted weight relationships to 1RM for multi-joint resistance exercises. For example, the estimate of 1RM from a sub-maximal performance at 5RM or 10RM, with R^2 ranging from 0.96 to 0.99, and SEE lower than 6 and 24 kg, respectively, for bench press and leg press, has been widely accepted as the alternative reference to predicted maximal muscle strength [30]. However, even when relying on submaximal muscle strength scores to estimate 1RM, it is well recognized that the same intrinsic determinant, such as sex and training status, can alter the maximum number of repetitions performed at certain fractions of 1RM [20]. Moreover, each type of exercise prescribed in resistance training requires its specific 1RM reference, and submaximal equations were not available to predict 1RM in different single- or multi-joint resistance exercises. Indeed, athletes should not agree to participate in time-consuming test procedures, or non-specific weight lifting, as these may disrupt their training planning.

However, the lack of a comparable sample of subjects to perform cross-validation of the present relationships hindered a better emphasis of the power of regional and whole-body FFM to predict lifting abilities in single- and multi-joint exercises because reproducibility and sensitivity were not evaluated. Nevertheless, the sample power for correlation analysis was above 80%, which is satisfactory to prevent type II errors. Moreover, cross-validation by applying the *PRESS* approach yielded values for R^2_p and SEE_p that were appropriate to strengthen the demonstrated correlations. In addition, the standardized 1RM protocol used in the current study may be a source of underestimation of the maximal strength during the attempt to attained the heaviest load in a single lifting [4]. Despite the possible underestimation of the actual maximal strength, this does not necessarily mean that a heavy load was not attained during the last lifting attempt, and the attained load was therefore ensured to be very close to the maximal one (i.e., >95% 1RM). Nonetheless, the results should, strictly, be applied to the management of 1RM values in subjects who met the following conditions: (a) expertise in the resistance exercise performance mode; (b) engagement in resistance training for at least two years; and (c) UL-FFM, WB-FFM, and arm cross-sectional area as adjustments to the observed correlation values.

5. Conclusions

The current findings evidenced the role of regional fat-free tissue for monitoring the muscle strength development in specific body regions. This demonstrated that regional FFM may be applied to parametrize muscle strength in different resistance exercises for upper and lower limbs, and would explain rates of 81% and 75% for single-joint exercises, respectively. As a suggestion to improve the reliance in these or other indices of regional and whole-body composition, future analysis should focus on how maximal weight lifting relates to fat-free tissue across randomized trials for both sexes, before and after intervention with resistance exercises planned for muscle strength improvements in single- and multi-joint exercises separately.

Author Contributions: Conceptualization, D.A.M., M.C.E., E.A.C., C.M.N. and D.M.P.F.; methodology, D.A.M., A.G.M., T.A.F.A., E.A.C., D.C.P.F. and D.M.P.F.; formal analysis, D.A.M., A.G.M.,

T.A.F.A., M.C.E., F.J.S., E.A.C. and D.M.P.F.; investigation, D.A.M., A.G.M., T.A.F.A., M.C.E., F.J.S., E.A.C., D.C.P.F. and D.M.P.F.; supervision, M.C.E., F.J.S., E.A.C., C.M.N. and D.M.P.F.; data curation, D.A.M. and D.M.P.F.; writing—original draft preparation, D.A.M., A.G.M., M.C.E., F.J.S., D.C.P.F. and D.M.P.F.; writing—review and editing, D.A.M., A.G.M., T.A.F.A., M.C.E., F.J.S., E.A.C., C.M.N. and D.M.P.F.; Visualization, D.A.M., A.G.M., T.A.F.A., M.C.E., F.J.S., E.A.C., D.C.P.F., C.M.N. and D.M.P.F.; funding acquisition, A.G.M., T.A.F.A., M.C.E., F.J.S., E.A.C. and D.M.P.F. All authors have read and agreed to the published version of the manuscript.

Funding: The authors would like to thank São Paulo Research Foundation—FAPESP (PROCESS 2016/04544-3) and Coordenação de Aperfeiçoamento de Pessoal de Nível Superior—Brazil (CAPES—Finance Code 001) for the partial financial support. The collaboration of T.A.F.A. and E.A.C was possible thanks to the scholarships granted by the CAPES, in the scope of the Program CAPES-PrInt, process number 88887.310463/2018-00, Mobility number 88887.580265/2020-00 and International Cooperation Project number 88887.572557/2020-00. This research was also funded by Foundation for Science and Technology, I.P., Grant/Award Number UIDB/04748/2020.

Institutional Review Board Statement: The study considered the guidelines of the Declaration of Helsinki and was submitted to the Local Ethics Committee of the University (CAEE: 19824719.3.0000.5398).

Informed Consent Statement: Informed consent was obtained from all participants in the study.

Data Availability Statement: The data that support the findings of this study are available from the corresponding and last author (mario.espada@ese.ips.pt and dalton.pessoa-filho@unesp.br), upon reasonable request.

Acknowledgments: The authors would like to thank the team of Laboreh (human performance optimization laboratory) for the helpful participation in data sampling, as well as all participants in the University Social Program: Sport Square Gymnasium (PROEX–UNESP-2020).

Conflicts of Interest: The authors declare no conflict of interest.

References

1. Fry, A.C. The role of resistance exercise intensity on muscle fiber adaptations. *Sports Med.* **2004**, *34*, 663–679. [CrossRef]
2. Giessing, J.; Eichmann, B.; Steele, J.; Fisher, J. A comparison of low volume 'high-intensity-training' and high volume traditional resistance training methods on muscular performance, body composition, and subjective assessments of training. *Biol. Sport* **2016**, *33*, 241–249. [CrossRef] [PubMed]
3. Ciolac, E.G.; Rodrigues-da-Silva, J.M. Resistance training as a tool for preventing and treating musculoskeletal disorders. *Sports Med.* **2016**, *46*, 1239–1248. [CrossRef] [PubMed]
4. Jaric, S. Muscle strength testing: Use of normalisation for body size. *Sports Med.* **2002**, *32*, 615–631. [CrossRef] [PubMed]
5. Ratamess, A.; Alvar, A.B.; Evetoch, T.K.; Housh, T.J.; Kibler, W.B.; Kraemer, W.J.J.; Triplett, N.T. ACSM Position Stand: Progression models in resistance training for healthy adults. *Med. Sci. Sports Exerc.* **2009**, *41*, 687–708. [CrossRef]
6. Heyward, V.H. *Advanced Fitness Assessment Exercise Prescription*, 3rd ed.; Human Kinetics: Champaign, IL, USA, 1997.
7. Mayhew, J.L.; McCormick, T.P.; Piper, F.C.; Kurth, A.L.; Arnold, M.D. Relationships of body dimensions to strength performance in novice adolescent male powerlifters. *Pediatr. Exerc. Sci.* **1993**, *5*, 347–356. [CrossRef]
8. Mayhew, J.L.; Piper, F.C.; Ware, J.S. Anthropometric correlates with strength performance among resistance trained athletes. *J. Sports Med. Phys. Fit.* **1993**, *33*, 159–165.
9. Hortobágyi, T.; Katch, F.I.; Katch, V.L.; LaChance, P.F.; Behnke, A.R. Relationships of body size, segmental dimensions, and ponderal equivalents to muscular strength in high-strength and low-strength subjects. *Int. J. Sports Med.* **1990**, *11*, 349–356. [CrossRef]
10. Hetzler, R.K.; Schroeder, B.L.; Wages, J.J.; Stickley, C.D.; Kimura, I.F. Anthropometry increases 1 repetition maximum predictive ability of NFL-225 test for Division IA college football players. *J. Strength Cond. Res.* **2010**, *24*, 1429–1439. [CrossRef]
11. Mayhew, J.L.; Ball, T.E.; Ward, T.E.; Hart, C.L.; Arnold, M.D. Relationship of structural dimensions to bench press strength in college males. *J. Sports Med. Phys. Fit.* **1991**, *31*, 135–141.
12. Cummings, B.; Finn, K.J. Estimation of a one repetition maximum bench press for untrained women. *J. Strength Cond. Res.* **1998**, *12*, 262–265.
13. Scanlan, J.M.; Ballmann, K.L.; Mayhew, J.L.; Lantz, C.D. Anthropometric dimensions to predict 1-RM bench press in untrained females. *J. Sports Med. Phys. Fit.* **1999**, *39*, 54–60.
14. Wood, L.E.; Dixon, S.; Grant, C.; Armstrong, N. Elbow flexion and extension strength relative to body or muscle size in children. *Med. Sci. Sports Exerc.* **2004**, *36*, 1977–1984. [CrossRef]
15. Materko, W.; Neves, C.E.B.; Santos, E.L. Prediction model of a maximal repetition (1RM) based on male and female anthropometrical characteristics. *Braz. J. Sports Med.* **2007**, *13*, 27–32. [CrossRef]

16. Pereira, M.I.R.; Gomes, P.S.C. Muscular strength and endurance tests: Reliability and prediction of one repetition maximum—Review and new evidences. *Braz. J. Sports Med.* **2003**, *9*, 336–346. [CrossRef]
17. Lemmer, J.T.; Martel, G.F.; Hurlbut, D.E.; Hurley, B.F. Age and sex differentially affect regional changes in one repetition maximum strength. *J. Strength Cond. Res.* **2007**, *21*, 731–737. [CrossRef]
18. Nindl, B.C.; Harman, E.A.; Marx, J.O.; Gotshalk, L.A.; Frykman, P.N.; Lammi, E.; Palmer, C.; Kraemer, W.J. Regional body composition changes in women after 6 months of periodized physical training. *J. Appl. Physiol.* **2000**, *88*, 2251–2259. [CrossRef]
19. Ramírez-Campillo, R.R.; Andrade, D.C.; Jara, C.C.; Olguín, C.H.; Lepin, C.A.; Izquierdo, M. Regional fat changes induced by localized muscle endurance resistance training. *J. Strength Cond. Res.* **2013**, *27*, 2219–2224. [CrossRef]
20. Fleck, S.J.; Mattie, C.; Martensen, L.H.C. Effect of resistance and aerobic training on regional body composition in previously recreationally trained middle-aged women. *Appl. Physiol. Nutr. Metab.* **2006**, *31*, 261–270. [CrossRef]
21. Pimenta, L.D.; Massini, D.A.; Santos, D.D.; Vasconcelos, C.M.T.; Simionato, A.R.; Gomes, L.A.T.; Pessôa, D.M. Bone health, muscle strength and fat-free mass: Relationships and exercise recommendations. *Rev. Bra. Med. Esporte* **2019**, *25*, 245–251. [CrossRef]
22. Nana, A.; Slater, G.J.; Hopkins, W.G.; Burke, L.M. Techniques for undertaking dual-energy X-ray absorptiometry whole-body scans to estimate body composition in tall and/or broad subjects. *Int. J. Sport Nut. Exerc. Metabol.* **2012**, *22*, 313–322. [CrossRef] [PubMed]
23. Mayhew, J.L.; Prinster, J.L.; Ware, J.S.; Zimmer, D.L.; Arabas, J.R.; Bemben, M.G. Muscular endurance repetitions to predict bench press in men of different training levels. *J. Sports Med. Phys. Fit.* **1995**, *35*, 108–113.
24. Baechle, T.R.; Earle, R.W. *Weight Training: Steps to Success*, 4th ed.; Human Kinetics: Champaign, IL, USA, 2012.
25. Ferguson, C.J. An Effect size primer: A guide for clinicians and researchers. professional psychology. *Res. Pract.* **2009**, *40*, 532–538. [CrossRef]
26. Díaz, S.P.; Fernández, S.P. Determinación del tamaño muestral para calcular la significación del coeficiente de correlación lineal. Unidad de Epidemiología Clínica y Bioestadística. Complexo Hospitalario Juan Canalejo. *A Coruña (España) Cad Aten Primaria* **2002**, *9*, 209–211.
27. Holiday, D.B.; Ballard, J.E.; McKeown, B.C. PRESS-related statistics: Regression tools for cross-validation and case diagnostics. *Med. Sci. Sports Exerc.* **1995**, *27*, 612–620. [CrossRef]
28. Malek, M.H.; Housh, T.J.; Berger, D.E.; Coburn, J.W.; Beck, T.W. A new non-exercise-based Vo_2max prediction equation for aerobically trained men. *J. Strength Cond. Res.* **2005**, *19*, 559–565. [CrossRef]
29. Braith, R.W.; Graves, J.E.; Leggett, S.H.; Pollock, M.L. Effect of training on the relationship between maximal and submaximal strength. *Med. Sci. Sports Exerc.* **1993**, *25*, 132–138. [CrossRef]
30. Reynolds, J.M.; Gordon, T.J.; Robergs, R.A. Prediction of one repetition maximum strength from multiple repetition maximum testing and anthropometry. *J. Strength Cond. Res.* **2006**, *20*, 584–592. [CrossRef]
31. Abdul-Hameed, U.; Rangra, P.; Shareef, M.Y.; Hussain, M.E. Reliability of 1-repetition maximum estimation for upper and lower body muscular strength measurement in untrained middle-aged Type 2 Diabetic patients. *Asian J. Sports Med.* **2012**, *3*, 267–273. [CrossRef]
32. Dohoney, P.; Chromiak, J.A.; Lemire, D.; Abadie, B.R.; Kovacs, C. Prediction of one repetition maximum (1-RM) strength from a 4–6 RM and a 7–10 RM submaximal strength test in healthy young adult males. *J. Exerc. Physiol.* **2012**, *5*, 54–59.
33. Donges, C.E.; Duffield, R. Effects of resistance or aerobic exercise training on total and regional body composition in sedentary overweight middle-aged adults. *Appl. Physiol. Nutr. Metab.* **2012**, *37*, 499–509. [CrossRef] [PubMed]
34. Abdelmoula, A.; Martin, V.; Bouchant, A.; Walrand, S.; Lavet, C.; Taillardat, M.; Maffiuletti, N.A.; Boisseau, N.; Duché, P.; Ratel, S. Knee extension strength in obese and nonobese male adolescents. *Appl. Physiol. Nutr. Metab.* **2012**, *37*, 269–275. [CrossRef] [PubMed]

International Journal of
***Environmental Research
and Public Health***

Article

Body Composition and Bioelectrical-Impedance-Analysis-Derived Raw Variables in Pole Dancers

Giada Ballarin [1,2], Luca Scalfi [2], Fabiana Monfrecola [2], Paola Alicante [2], Alessandro Bianco [2], Maurizio Marra [3] and Anna Maria Sacco [2,*]

[1] Department of Movement Sciences and Wellbeing, University of Naples "Parthenope", 80133 Naples, Italy; giada.ballarin001@studenti.uniparthenope.it
[2] Department of Public Health, School of Medicine, Federico II University, 80131 Naples, Italy; scalfi@unina.it (L.S.); fabiana.monfrecola@hotmail.it (F.M.); paola.alicante@unina.it (P.A.); alessandrobianco92@gmail.com (A.B.)
[3] Department of Clinical Medicine and Surgery, School of Medicine, Federico II University, 80131 Naples, Italy; marra@unina.it
* Correspondence: annamaria.sacco@unina.it

Abstract: Few data are available on the body composition of pole dancers. Bioelectrical impedance analysis (BIA) is a method that is used to estimate fat-free mass (FFM) and fat mass (FM), while raw BIA variables, such as the impedance ratio (IR) and phase angle (PhA), are markers of body cell mass and the ratio between extracellular and total body water. The aim of this study was to evaluate the body composition of pole dancers compared to controls, in particular, those raw BIA variables that are considered as markers of muscle composition. Forty female pole dancers and 59 controls participated in the study. BIA was performed on the whole body and upper and lower limbs, separately, at 5, 50, 100 and 250 kHz. The FFM, FFM index, FM and body fat percentage (BF%) were predicted. The bioelectrical impedance indexes IR and PhA were also considered. Pole dancers exhibited higher FFMI and BI indexes and lower BF%. PhA was greater and IRs were smaller in pole dancers than in controls for the whole body and upper limbs. Considering the training level, FFM, whole-body IR and PhA were higher in the professionals than non-professionals. Raw BIA variables significantly differed between the pole dancers and controls, suggesting a higher BCM; furthermore, practicing pole dancing was associated with a greater FFM and lower FM.

Keywords: bioelectrical impedance analysis; muscle composition; phase angle; impedance ratio; pole dance

Citation: Ballarin, G.; Scalfi, L.; Monfrecola, F.; Alicante, P.; Bianco, A.; Marra, M.; Sacco, A.M. Body Composition and Bioelectrical-Impedance-Analysis-Derived Raw Variables in Pole Dancers. *Int. J. Environ. Res. Public Health* **2021**, *18*, 12638. https://doi.org/10.3390/ijerph182312638

Academic Editors: Rafael Oliveira and João Paulo Brito

Received: 15 October 2021
Accepted: 25 November 2021
Published: 30 November 2021

Publisher's Note: MDPI stays neutral with regard to jurisdictional claims in published maps and institutional affiliations.

1. Introduction

The evaluation of body composition is crucial not only for assessing nutritional status in the general population but also for athletes for the monitoring of training and performance.

Anthropometry and bioelectrical impedance analysis (BIA) are both field methods that are widely used to assess the body composition of athletes [1]. In particular, BIA is a simple, non-invasive technique that measures the electrical characteristics of the human body, i.e., impedance (Z) and phase angle (PhA) (from those, resistance (R) and reactance (Xc) can also be derived). Total body water (TBW), fat-free mass (FFM) and fat mass (FM) can be estimated by means of predictive equations that include BIA variables and very often other variables, such as age, height and body mass; some equations were specifically developed for athletes. Since these specific equations [2–4] have not been definitively validated, the BIA-derived estimation of body composition should be considered with caution. In particular, the BIA method has an error of 4–8% compared to criterion methods, which could be even more evident in athletes [3]. On the other hand, the BIA estimates of body composition might give some interesting evidence on body composition on a groupwise basis. Of note, the bioimpedance index at 50 kHz (BI index = height^2/Z at 50 kHz) is commonly considered as a logical predictor of FFM and TBW [5,6]. Finally, it should be noted that BIA may be performed on the whole body but also separately for upper limbs and lower limbs (segmental BIA), giving, at least in theory, the chance for evaluating appendicular muscle mass [7–9].

Raw BIA variables, such as the impedance ratio (IR), which is the ratio between Z at high frequencies and Z at low frequencies, and PhA at 50 kHz, are those that are directly measured by an analyzer. Their assessment in sportspeople is motivated by the fact that IR and PhA may be considered as potential markers of both body cell mass (BCM) and the ratio between extracellular water and total body water (ECW/TBW ratio) [10–13]; in other words, both these variables may give information on the electrical properties, as well as the FFM composition and/or muscle composition. IR and PhA were related to muscle strength and physical activity [14,15] as well, while in the first decades of life and elderly people PhA was associated with muscle performance [16,17], isolated or grouped physical fitness indicators [18,19], and cardiorespiratory fitness [20]. As reported in a recent systematic review [21], it is still to be determined to what extent PhA differs between different sports and due to training/untraining; some studies showed that mean whole-body PhA is higher in athletes vs. controls [21,22], while, to the best of our knowledge, so far no data are available on IRs in sportspeople and only limited data exists on segmental BIA [7–9,21,22].

With regard to sports activities, pole dancing is a type of functional training that involves the use of a vertical pole to perform exercises and figures. A training session, called a pole class, lasts between 60 and 90 min (possibly depending on training level) and can be subdivided into three parts: warm-up and strengthening exercises are performed first; then the specific tool figures are studied, with increasing difficulty of execution, while cooldown exercises close the session. Pole dancing may be considered a moderate-intensity cardiorespiratory endurance exercise that, if practiced regularly, leads to a significant increase in aerobic capacity, resistance, flexibility, and motor coordination [23,24].

To the best of our knowledge, only a single study has evaluated the body composition of female pole dancers, attributing an increase in postural strength and stability to the more experienced athletes, but no changes in body composition [25]. Looking at similar sports, rhythmic gymnasts exhibited lower body mass, body mass index (BMI) and skinfold thickness compared to other athletes [26], while gymnasts had a reduced body fat percentage (BF%) compared to controls with the same BMI [27,28]. Dancers had similar BF% but higher levels of FFM and muscle mass than controls, whereas low values of FFM and fat mass (FM) were observed in the case of underweight athletes [29]. Finally, in sedentary women, a choreographed fitness group workout contributed to reducing FM and increasing muscle mass [30].

Against this background, the aim of this cross-sectional study was to evaluate the body composition of pole dancers (non-professional and professional athletes) compared to controls, with a particular interest in the raw BIA variables that are thought to be markers of FFM composition and/or muscle composition. In addition, a segmental BIA evaluation was performed to explore the electrical characteristics of upper or lower limbs.

2. Methods

Forty female pole dancers and fifty-nine control young women participated in the study. Pole dancers were recruited from among those going to two gyms in Naples (a participation rate of 89%) and were non-professional performers (hereafter defined as amateurs) ($n = 33$), who trained 2–4 h a week in two sessions (18–36 months of specific training), and professionals ($n = 7$) who were pole dance trainers (at least 60 months and more than 6 h a week of specific training). Controls ($n = 59$) were sedentary women (at most 1 h of physical training twice a week) and were recruited from among the female students attending the "Federico II" University of Naples. All subjects were healthy. The Ethics Committee of the "Federico II" University of Naples approved the research protocol and subjects gave their informed consent to participate in the study.

The participants avoided physical exercise for 24 h before the measurement session and were studied by the same operator following standard procedures. Data were collected between March and April 2019 in four sessions for pole dancers and six sessions for controls (data on $\geq$10 women were collected in each session). The general schedule was similar in the two groups of pole dancers, with different intensities of training programs based on their training level.

Body mass was measured to the nearest 0.1 kg using a platform beam scale and height was measured to the nearest 0.5 cm using a stadiometer (Seca, Hamburg, Germany). Participants were asked to remove shoes and heavy clothes prior to weighing. BMI was then calculated as body mass (kg)/height^2 (m^2).

Height was measured according to standard procedures. The participants were asked to stand up straight against the backboard with their body weight evenly distributed and both feet flat on the stadiometer platform, while the head was in the Frankfort horizontal plane [31].

Mid-arm circumference and triceps skinfold thickness (Holtain skinfold caliper) were measured on both body sides and, subsequently, the arm muscle area (AMA), corrected for the bone area, and arm fat area (AFA) were calculated as follows [32]:

$$AMA = [(Mid\text{-}arm\ Circumference - \pi \times TSF) \times 2/4\pi] - 6.5$$

$$AFA = Arm\ total\ area - AMA$$

BIA was performed using a HUMAN IM TOUCH multi-frequency analyzer (DS MEDICA, Milan, Italy) in standardized conditions: ambient temperature between 23–25 °C, fast for >3 h, empty bladder and supine position for 10 min. Data on Z at four different frequencies (5, 50, 100 and 250 kHz) and PhA at 50 kHz were considered for the statistical analysis. Precision resistors and capacitors (reference electronic circuits) were routinely used for calibration. The reproducibility of the BIA was previously assessed in ten healthy volunteers on subsequent days with a mean coefficient of variation of 1.5% for Z (at each of the different frequencies considered) and 2% for the phase angle at 50 kHz.

The 250 kHz/5 kHz IR may be used as a proxy marker of fluid distribution and was recently related by our group to mortality in patients with chronic obstructive pulmonary disease [10,14]. Subjects were asked to lie down with their legs and arms slightly abducted (~30°) to ensure no contact between body segments. The measuring electrodes were placed on the anterior surface of the wrist and ankle, and the injecting electrodes were placed on the dorsal surface of the hand and the foot, respectively [13]. Segmental BIA was performed using a six-electrode technique according to Organ [33].

Whole-body BI indexes were calculated as height^2 divided by Z as markers of ECW (Z at a low frequency of 5 kHz) and FFM (Z at high frequencies of 50, 100 or 250 kHz). In addition, two other raw variables were measured for the whole body and upper or lower limbs separately: (1) IR is commonly calculated as the ratio between Z at 200, 250 or 300 kHz and Z at 5 kHz [10]. In the present study, data were obtained for three ratios: Z 50 kHz/Z 5 kHz (IR50/5), Z 100 kHz/Z 5 kHz (IR100/5), and Z 250 kHz/Z 5 kHz (IR250/5). (2) PhA was measured at 50 kHz, as described in the literature. To the best of our knowledge, there has been little interest in applied physiology and human nutrition for evaluating the phase angle at frequencies other than 50 kHz. In all cases, mean values for the dominant (D) and non-dominant (ND) body sides were considered for statistical analysis to give more consistent results for the entire body. FFM was estimated using the Sun equation [34], which is a well-known equation that was proposed for the general population aged 12–94 years and which is also expected to perform well in young women with a higher physical activity level but no very major changes in body composition.

Whole-body FFM was calculated as follows:

$$FFM = -9.53 + 0.69 \times height^2/resistance + 0.17 \times body\ mass + 0.02 \times resistance$$

where the resistance at 50 kHz was derived by multiplying Z by the cosine of PhA.

Finally, FM was obtained from the difference between body mass and FFM, while the fat-free mass index (FFMI) was calculated as FFM (kg)/height^2 (m^2).

Statistical Analysis

Data obtained during the routine examination of athletes or control subjects were retrospectively retrieved. With a type I error rate of 0.05 and a type II error rate of 0.20, a sample size of 85 subjects is required to determine whether a correlation coefficient of 0.3 differs from zero.

Results are expressed as mean $\pm$ standard deviation (with some exceptions, see below). Statistical significance was pre-determined as $p < 0.05$. Effect size was calculated according to Cohen [35].

All statistical analyses were carried out using the Statistical Package for Social Sciences (SPSS Inc., Chicago, IL, USA) version 26. One-way analysis of variance was performed to assess the differences between two groups (pole dancers vs. controls or amateurs vs. professionals). Partial correlation was used to assess the relationships between the variables. The general linear model (GLM) was used to assess how several variables affected the continuous variables. From a practical point of view, it was used to compare the body composition between groups after controlling for body mass; adjusted means $\pm$ standard errors were provided by this statistical procedure.

3. Results

The general characteristics of the study groups are summarized in Table 1. Despite no difference in body mass and BMI, the pole dancers exhibited lower BF% compared to the controls (-14%). Correspondingly, the AMA was significantly greater and the AFA was smaller in the pole dance than in the control group (Table 1).

Table 1. Individual characteristics and body composition in female pole dancers and controls.

	Pole Dancers (*n* = 40)	Controls (*n* = 59)	*p*-Value	Cohen's d
Age (years)	27.4 $\pm$ 5.1	26.8 $\pm$ 4.7	0.561	0.12
Body mass (kg)	57.0 $\pm$ 6.9	58.6 $\pm$ 6.4	0.225	0.24
Height (cm)	160.3 $\pm$ 5.1	161.9 $\pm$ 4.9	0.139	0.32
BMI (kg/m^2)	22.2 $\pm$ 2.3	22.3 $\pm$ 1.8	0.747	0.05
Fat-free mass, FFM (kg)	43.5 $\pm$ 3.5	43.0 $\pm$ 3.1	0.448	0.15
Fat-free mass index, FFMI (kg/m^2)	16.9 $\pm$ 1.1	16.4 $\pm$ 0.8	0.007 *	0.52
Fat mass, FM (kg)	13.5 $\pm$ 4.3	15.6 $\pm$ 4.1	0.013 *	0.50
Percentage body fat, BF%	23.2 $\pm$ 4.7	26.3 $\pm$ 4.4	0.001 *	0.74
Arm muscle area D, AMA (cm^2)	52.5 $\pm$ 9.4	48.9 $\pm$ 8.9	0.060	0.39
Arm fat area D, AFA (cm^2)	2.0 $\pm$ 0.5	2.2 $\pm$ 0.8	0.047 *	0.30
Arm muscle area ND, AMA (cm^2)	51.8 $\pm$ 10.4	48.0 $\pm$ 8.4	0.045 *	0.40
Arm fat area ND, AMA (cm^2)	2.0 $\pm$ 0.6	2.2 $\pm$ 0.7	0.098	0.31

Data are expressed as mean $\pm$ standard deviation. * $p < 0.05$. BMI—body mass index. FFM and FM were estimated from the BIA; AMA was corrected for bone area. D—dominant side and ND—non-dominant side of the body. Effect size: Cohen's d $\leq$ 0.2 = small, 0.2 < d $\leq$ 0.5 = small to medium, 0.5 < d $\leq$ 0.8 = medium to large, d > 0.8 = large.

As for the raw BIA variables, the whole-body and upper limb Z values were lower in the pole dancers than in the controls; for instance, Z at 250 kHz was 485 $\pm$ 50 vs. 519 $\pm$ 38 kHz and 240 $\pm$ 28 vs. 271 $\pm$ 20 kHz, respectively (d = 0.39 and d = 0.72; $p < 0.001$), with small differences (<2%) between the D and ND body side. Furthermore, Table 2 indicates that the BI indexes at 5, 50, 100 and 250 kHz were higher in the pole dancers than in the controls (+4.3, +4.9, +5.3 and +5.3%, respectively). These differences in the mean values of different Z and BI indexes persisted after adjusting for age and mass (data not shown). After controlling for groups, a partial correlation indicated that whole-body BI indexes were associated with AMA (r > 0.450 for 50, 100 and 250 kHz vs. r = 0.416 for 5 kHz) but not with AFA.

Table 2. Bioimpedance indexes, impedance ratios and phase angles that were measured for the whole body and upper and lower limbs in female pole dancers and controls.

		Pole Dancers ($n = 40$)	Controls ($n = 59$)	p-Value	Cohen's d
Bioimpedance Index (Ω)					
Whole body	5 kHz	41.1 ± 4.2	39.4 ± 3.8	0.043 *	0.42
	50 kHz	46.8 ± 4.9	44.6 ± 4.1	0.018 *	0.49
	100 kHz	49.7 ± 5.3	47.2 ± 4.3	0.013 *	0.52
	250 kHz	53.5 ± 5.7	50.8 ± 4.6	0.011 *	0.52
Impedance Ratio					
Whole body	Z 50/Z 5 kHz	0.878 ± 0.014	0.883 ± 0.014	0.060	0.36
	Z 100/Z 5 kHz	0.827 ± 0.017	0.835 ± 0.017	0.039 *	0.47
	Z 250/Z 5 kHz	0.768 ± 0.018	0.775 ± 0.018	0.058	0.39
Upper limbs	Z 50/Z 5 kHz	0.887 ± 0.013	0.897 ± 0.015	<0.001 *	0.71
	Z 100/Z 5 kHz	0.837 ± 0.016	0.852 ± 0.018	<0.001 *	0.88
	Z 250/Z 5 kHz	0.769 ± 0.019	0.783 ± 0.020	<0.001 *	0.72
Lower limbs	Z 50/Z 5 kHz	0.867 ± 0.018	0.865 ± 0.018	0.451	0.13
	Z 100/Z 5 kHz	0.816 ± 0.022	0.814 ± 0.021	0.718	0.09
	Z 250/Z 5 kHz	0.771 ± 0.025	0.769 ± 0.024	0.765	0.08
Phase Angle ($°$)					
	Whole body	6.07 ± 0.56	5.85 ± 0.56	0.063	0.39
	Upper limbs	5.27 ± 0.59	4.76 ± 0.56	<0.001 *	0.89
	Lower limbs	7.05 ± 0.70	7.06 ± 0.69	0.974	0.01

Data are expressed as mean $\pm$ standard deviation. * $p < 0.05$. BI index—bioimpedance index (height^2/Z), IR—impedance ratio, PhA—phase angle. Cohen's d ≤ 0.2—small, $0.2 < d \leq 0.5$—small to medium, $0.5 < d \leq 0.8$—medium to large, d > 0.8—large.

As shown in Table 2, PhA was greater in pole dancers than in controls by 3.8% for the whole body (d = 0.39 and $p = 0.063$) and by 10.7% for upper limbs (d = 0.89 and $p < 0.001$), whereas there was no difference for lower limbs. IRs were lower in the pole dance group than in the control group, again more markedly for upper limbs (Table 2). The differences for upper limbs were still found in both cases even after controlling for age and body mass. In particular, multiple regression analysis indicated age and body mass as predictors of IR250/5 (for the whole model: $R^2 = 0.117$, $F(2,87) = 6.83$, $p = 0.002$) and PhA ($R^2 = 0.053$, $F(2,87) = 5.90$, $p = 0.017$). Of note, no relationships were detected between IRs or PhA and body composition.

There was no significant association of PhA or IRs with height, mass, BMI, FFM, FM, AMA or BI indexes ($p > 0.20$, data not shown). On the other hand, after adjusting for groups, a partial correlation indicated a moderate association between the upper limb and lower limb values of PhA (r = 0.463), IR50/5 (r = 0.538), IR100/5 (r = 0.531) and IR250/5 (r = 0.514).

With respect to the training level, professional and amateur pole dancers did not differ in terms of body mass (55.6 ± 4.2 vs. 57.3 ± 7.3 kg) and BMI (22.0 ± 2.3 vs. 22.2 ± 2.3 kg/m^2). The GLM indicated that, after adjusting for body mass, FFM (mean $\pm$ SEM, 45.3 ± 0.6 vs. 43.7 ± 0.3 kg, $p = 0.024$) was greater in the more trained than in the less trained athletes, while BF% was smaller (21.4 ± 11.1 vs. $24.2 \pm 0.5\%$, $p = 0.023$, respectively). In particular, multiple regression analysis was used to test whether training level and body mass significantly predicted participants' FFM and BF%. The results indicated that the two predictors explained 75% of the total variance for FFM ($R^2 = 0.75$, F(2,86) = 130.9, $p < 0.001$) and 64% of the total variance for BF% ($R^2 = 0.64$, F(2,86) = 78.0, $p < 0.001$).

Turning to raw BIA variables, whole-body PhA and IRs were higher, but not significantly (d between 0.5 and 0.8; p between 0.05 and 0.10), in the professional athletes than in the amateur athletes (Table 3). More evident differences (Figure 1) emerged for the upper limbs: the professional pole dancers had significantly smaller IRs and greater PhA than the amateur athletes and controls, and the same was true when amateurs were compared to the controls (d = 0.99 and $p < 0.05$). After taking into consideration the training level as a predictor, no significant relationships were found between IRs or PhA and body mass or body composition.

Table 3. Bioimpedance index, impedance ratio and phase angle measured for the whole body and upper and lower limbs in amateur and professional pole dancers.

		Professional Pole Dancers ($n = 7$)	Amateur Pole Dancers ($n = 33$)	p-Value	Cohen's d
Impedance Ratio					
Whole body	Z 50/Z 5 kHz	0.869 ± 0.015	0.879 ± 0.014	0.079	0.70
	Z 100/Z 5 kHz	0.817 ± 0.018	0.830 ± 0.016	0.072	0.76
	Z 250/Z 5 kHz	0.756 ± 0.021	0.771 ± 0.018	0.058	0.77
Upper limbs	Z 50/Z 5 kHz	0.875 ± 0.010	0.889 ± 0.013	<0.001 *	1.2
	Z 100/Z 5 kHz	0.824 ± 0.014	0.840 ± 0.015	<0.001 *	1.1
	Z 250/Z 5 kHz	0.753 ± 0.018	0.772 ± 0.018	<0.001 *	1.1
Lower limbs	Z 50/Z 5 kHz	0.863 ± 0.021	0.868 ± 0.018	0.463	0.26
	Z 100/Z 5 kHz	0.810 ± 0.025	0.817 ± 0.021	0.435	0.30
	Z 250/Z 5 kHz	0.763 ± 0.030	0.772 ± 0.025	0.355	0.33
Phase Angle (°)					
	Whole body	6.37 ± 0.57	6.00 ± 0.55	0.117	0.66
	Upper limbs	5.66 ± 0.56	5.19 ± 0.56	0.041 *	0.99
	Lower limbs	7.11 ± 0.80	7.04 ± 0.70	0.821	0.09

Data are expressed as mean ± standard deviation. BI index—bioimpedance index calculated as $height^2/Z$. * $p < 0.05$. Cohen's d ≤ 0.2—small, 0.2 < d ≤ 0.5—small to medium, 0.5 < d ≤ 0.8—medium to large, d > 0.8—large.

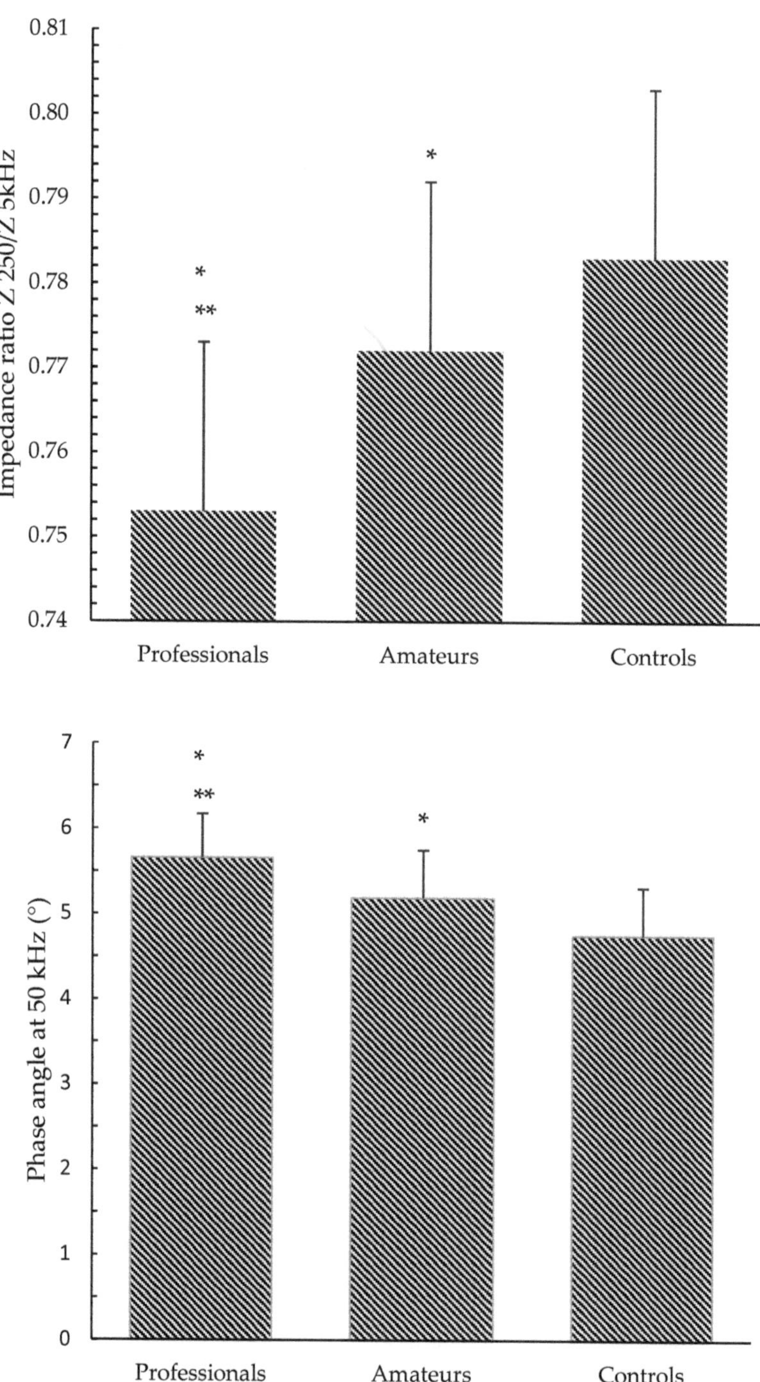

Figure 1. Impedance ratio Z 250 kHz/Z 5 kHz and phase angle at 50 kHz in amateur or professional pole dancers compared to control women. * $p < 0.05$ vs. controls ** $p < 0.05$ vs. amateurs and controls.

4. Discussion

In the present study, raw BIA variables that may be considered as markers of FFM composition and/or muscle composition significantly varied between female pole dancers and controls, showing different electrical characteristics of the body and suggesting higher BCM; in addition, pole dancers exhibited lower BIA-derived FM and BF%.

We performed a cross-sectional study on a relatively large group of pole dancers compared to sedentary controls, bearing in mind that the effects of this type of training on body composition have so far been poorly explored [25]. Unfortunately, there was no information regarding participants' body composition before starting the training. Indeed, in light of the difficulties in carrying out long-term intervention studies, the present cross-sectional study is expected to provide some preliminary insights regarding the effect of pole dancing on body composition.

Body composition was assessed using BIA, which is a technique that is widely used in athletes [1]. Since the specific equations developed for athletes [2–4] have not been definitively validated [3,13], BIA-derived estimation of body composition should be considered with caution. In particular, the BIA method has an error of 4–8% compared to criterion methods, which could be even more evident in athletes [3]. On the other hand, the BIA estimates of body composition might give some evidence on body composition on a groupwise basis. In the present study, the Sun equation was chosen to predict FFM [34]; this formula was developed in a large sample of healthy subjects using a multicomponent model, it is widely used, and it is expected to also perform well in young women with a higher physical activity level but no major changes in body composition.

Thus, we looked first at BIA-derived estimates of body compartments. Despite having similar body mass and BMI, pole dancers had lower FM and BF% compared to the controls. These findings are in agreement with those reported in previous cross-sectional studies that showed higher FFM and smaller FM in female gymnasts and dancers [26–28]. Of note, the study by Nawrocka et al. [25] on the body composition of pole dancers did not include a control group. Overall, our results suggest a significant, but small effect of pole dance training on body composition, with a moderate to high effect size for BF% (d = 0.74 and $p = 0.001$).

As an alternative approach, IRs and PhA (for the whole body and upper and lower limbs, separately) were directly (no predictive equations used) determined in pole dancers and controls as a qualitative approach to body composition analysis [13]. Both those raw BIA variables may be effective in exploring FFM composition and muscle composition in terms of the electrical characteristics of tissues, as well as BCM and the ECW/TBW ratio [10–13]. Interestingly, IRs and PhA have also been associated with muscle strength and physical activity [14,15,19]. A few cross-sectional studies showed that mean whole-body PhA is higher in athletes vs. controls, while, to the best of our knowledge, no data so far are available on IRs [21]; of note, a recent paper showed, as expected, a high correlation between IRs and PhA [19]. In addition, it is still to be determined to what extent IR and/or PhA may vary between different sports and with training/untraining [13,21]. Facing this background, although in our experience data on IR or PhA are very reproducible, the use of these BIA variables in longitudinal studies or single athletes should be better defined and considered with caution.

IR is commonly calculated as the ratio between Z at high frequency and Z at low frequency [10]. The ratio between Z at 200 kHz and Z at 5 kHz (IR200/5) is widely used but still not formally indicated as the only one to be taken into consideration. Results on three different IRs are reported here, with IR250/5 being very close to IR 200/5. The three IRs were all slightly smaller in the pole dance group compared to the control group. At first glance, these differences in IRs were small in percentage terms, but relevant when compared to the corresponding standard deviations. For instance, the difference in IR250/5 was 0.007, while the pooled standard deviation was 0.019 (d = 0.39 and $p = 0.058$). Regarding another raw BIA variable, whole-body PhA, which was measured at 50 kHz, as commonly described in the literature [10,21], was only slightly higher in pole dancers compared to controls (low size effect). Overall, only minor changes were observed for the whole body.

It is clear that segmental BIA, as performed on upper limbs and lower limbs separately, may give, at least in theory, the chance to evaluate appendicular muscle mass more directly [7–9]. Few previous papers have performed this type of measurement in athletes; they, for instance, showed greater PhA for both lower and upper limbs in female volleyball players compared to controls [8]. Our study yielded some results of interest: lower limb IRs and PhA did not differ between the groups, while a marked difference emerged for upper limbs (d = 0.72 and $p < 0.001$ and d = 0.89 and $p < 0.001$, respectively), suggesting some effects of pole dancing on different muscle groups. Of note, those differences persisted after adjusting for age plus body mass or plus body composition. Thus, segmental measurement seemed to be effective in detecting differences in raw BIA variables, which should be examined in detail by further studies that consider various types of training and use different criterion methods for assessing body composition.

Even if the interpretation of data on professional dancers (Table 3) should be discussed with caution due to the limited sample size, some stimulating findings emerged: compared to amateurs, they had lower IRs and higher PhA for the upper limbs, suggesting a possible relationship between workout volume and the electrical characteristics of muscle. In addition, smaller IRs and greater PhA for upper limbs were still observed in amateur athletes compared to the controls (Figure 1).

Athletes and controls were studied in standardized conditions by a single experienced operator, while BIA was performed on both body sides to ensure a more reliable assessment of the electrical characteristics of the body. A large proportion of the pole dancers going to two different gyms participated in the study, while control women were selected among those who were enrolled in a study on university students who did low amounts of physical activity.

Indeed, there are limitations to the study that should be considered. It was a single-center cross-sectional study in which body composition was evaluated by means of a field method. Furthermore, we specifically focused on the assessment of raw BIA variables, such as IR and PhA, that are markers of FFM composition or muscle composition and cannot easily be compared with a proper criterion technique. In addition, there was no information regarding participants' body composition before starting the training, and it was not possible to carry out a very accurate evaluation of the strengthening or conditioning workouts.

5. Conclusions

In conclusion, care must be taken not to overinterpret the results of the present study. The main findings were that raw BIA variables that may be considered as markers of FFM composition or muscle composition significantly differed between female pole dancers and controls, suggesting higher BCM, as well as a lower ECW/TBW ratio; in addition, practicing pole dancing is associated with lower FM and BF%.

Differences in PhA and IRs may suggest modifications in the electrical characteristics of the body that seem to be more marked for the upper limbs and possibly in professional than amateur athletes and that was similar for the three IRs considered. These findings are in line with the literature describing changes in raw BIA variables and body composition due to regular physical exercise [8,9,21,22]. Further studies, especially intervention studies, are needed to define the best approach to use BIA in order to measure raw BIA variables and possibly track changes in the body composition of athletes with time.

Author Contributions: G.B., F.M. and A.B. collected the data. G.B., P.A. and M.M. analyzed the data. G.B., L.S., A.M.S. and M.M. designed the study and wrote the manuscript. A.M.S. supervised the project. All authors discussed the results and commented on the manuscript. All authors have read and agreed to the published version of the manuscript.

Funding: This research received no external funding.

Institutional Review Board Statement: The Ethics Committee of the "Federico II" University of Naples (N: 42/17) approved the research protocol and the study was conducted in accordance with the Declaration of Helsinki.

Informed Consent Statement: All participants gave written informed consent prior to being enrolled in the study.

Data Availability Statement: The datasets used and/or analyzed during the current study are available from the corresponding author on reasonable request.

Acknowledgments: The authors would like to thank Giuseppe Abate for his technical assistance.

Conflicts of Interest: The authors declare no conflict of interest.

References

1. Ackland, W.P.T.R.; Lohman, T.G.; Sundgot-Borgen, J.; Maughan, R.J.; Meyer, N.L.; Stewart, A.; Müller, W. Current Status of Body Composition Assessment in Sport. *Sports Med.* **2012**, *42*, 227–249. [CrossRef]
2. Castizo-Olier, J.; Irurtia, A.; Jemni, M.; Carrasco-Marginet, M.; Fernández-García, R.; Rodríguez, F.A. Bioelectrical impedance vector analysis (BIVA) in sport and exercise: Systematic review and future perspectives. *PLoS ONE* **2018**, *13*, e0197957. [CrossRef]
3. Moon, J.R. Body composition in athletes and sports nutrition: An examination of the bioimpedance analysis technique. *Eur. J. Clin. Nutr.* **2013**, *67*, S54–S59. [CrossRef]
4. Sardinha, L.; Correia, I.R.; Magalhães, J.P.; Júdice, P.; Silva, A.M.; Hetherington-Rauth, M. Development and validation of BIA prediction equations of upper and lower limb lean soft tissue in athletes. *Eur. J. Clin. Nutr.* **2020**, *74*, 1646–1652. [CrossRef]
5. Kyle, U.G.; Bosaeus, I.; De Lorenzo, A.D.; Deurenberg, P.; Elia, M.; Gomez, J.M.; Heitmann, B.L.; Kent-Smith, L.; Melchior, J.-C.; Pirlich, M.; et al. Bioelectrical impedance analysis? Part I: Review of principles and methods. *Clin. Nutr.* **2004**, *23*, 1226–1243. [CrossRef]
6. Khalil, S.F.; Mohktar, M.S.; Ibrahim, F. The Theory and Fundamentals of Bioimpedance Analysis in Clinical Status Monitoring and Diagnosis of Diseases. *Sensors* **2014**, *14*, 10895–10928. [CrossRef]
7. Burdukiewicz, A.; Pietraszewska, J.; Andrzejewska, J.; Chromik, K.; Stachoń, A. Asymmetry of Musculature and Hand Grip Strength in Bodybuilders and Martial Artists. *Int. J. Environ. Res. Public Health* **2020**, *17*, 4695. [CrossRef] [PubMed]
8. Di Vincenzo, O.; Marra, M.; Sammarco, R.; Speranza, E.; Cioffi, I.; Scalfi, L. Body composition, segmental bioimpedance phase angle and muscular strength in professional volleyball players compared to a control group. *J. Sports Med. Phys. Fit.* **2020**, *60*, 870–874. [CrossRef]
9. Marra, M.; Da Prat, B.; Montagnese, C.; Caldara, A.; Sammarco, R.; Pasanisi, F.; Corsetti, R. Segmental bioimpedance analysis in professional cyclists during a three week stage race. *Physiol. Meas.* **2016**, *37*, 1035–1040. [CrossRef] [PubMed]
10. Lukaski, H.C.; Kyle, U.G.; Kondrup, J. Assessment of adult malnutrition and prognosis with bioelectrical impedance analysis: Phase Angle and Impedance Ratio. *Curr. Opin. Clin. Nutr. Metab. Care* **2017**, *20*, 330–339. [CrossRef] [PubMed]
11. Francisco, R.; Matias, C.N.; Santos, D.A.; Campa, F.; Minderico, C.S.; Rocha, P.; Heymsfield, S.B.; Lukaski, H.; Sardinha, L.B.; Silva, A.M. The Predictive Role of Raw Bioelectrical Impedance Parameters in Water Compartments and Fluid Distribution Assessed by Dilution Techniques in Athletes. *Int. J. Environ. Res. Public Health* **2020**, *17*, 759. [CrossRef]
12. Silva, A.M.; Nunes, C.L.; Matias, C.N.; Rocha, P.M.; Minderico, C.S.; Heymsfield, S.B.; Lukaski, H.; Sardinha, L. Usefulness of raw bioelectrical impedance parameters in tracking fluid shifts in judo athletes. *Eur. J. Sport Sci.* **2019**, *20*, 734–743. [CrossRef]
13. Campa, F.; Toselli, S.; Mazzilli, M.; Gobbo, L.A.; Coratella, G. Assessment of Body Composition in Athletes: A Narrative Review of Available Methods with Special Reference to Quantitative and Qualitative Bioimpedance Analysis. *Nutrients* **2021**, *13*, 1620. [CrossRef] [PubMed]
14. de Blasio, F.; Scalfi, L.; Di Gregorio, A.; Alicante, P.; Bianco, A.; Tantucci, C.; Bellofiore, B. Raw Bioelectrical Impedance Analysis Variables Are Independent Predictors of Early All-Cause Mortality in Patients With COPD. *Chest* **2019**, *155*, 1148–1157. [CrossRef] [PubMed]
15. Mundstock, E.; Amaral, M.A.; Baptista, R.R.; Sarria, E.E.; dos Santos, R.R.G.; Filho, A.D.; Rodrigues, C.A.S.; Forte, G.C.; Castro, L.; Padoin, A.V.; et al. Association between phase angle from bioelectrical impedance analysis and level of physical activity: Systematic review and meta-analysis. *Clin. Nutr.* **2019**, *38*, 1504–1510. [CrossRef]
16. Yamada, Y.; Itoi, A.; Yoshida, T.; Nakagata, T.; Yokoyama, K.; Fujita, H.; Kimura, M.; Miyachi, M. Association of bioelectrical phase angle with aerobic capacity, complex gait ability and total fitness score in older adults. *Exp. Gerontol.* **2021**, *150*, 111350. [CrossRef] [PubMed]
17. Hetherington-Rauth, M.; Baptista, F.; Sardinha, L. BIA-assessed cellular hydration and muscle performance in youth, adults, and older adults. *Clin. Nutr.* **2020**, *39*, 2624–2630. [CrossRef] [PubMed]
18. Martins, P.C.; Moraes, M.S.; Silva, D.A.S. Cell integrity indicators assessed by bioelectrical impedance: A systematic review of studies involving athletes. *J. Bodyw. Mov. Ther.* **2020**, *24*, 154–164. [CrossRef] [PubMed]
19. Sacco, A.M.; Valerio, G.; Alicante, P.; Di Gregorio, A.; Spera, R.; Ballarin, G.; Scalfi, L. Raw bioelectrical impedance analysis variables (phase angle and impedance ratio) are significant predictors of hand grip strength in adolescents and young adults. *Nutrition* **2021**, *91–92*, 111445. [CrossRef] [PubMed]

20. Langer, R.D.; da Costa, K.G.; Bortolotti, H.; Fernandes, G.A.; de Jesus, R.S.; Gonçalves, E.M. Phase angle is associated with cardiorespiratory fitness and body composition in children aged between 9 and 11 years. *Physiol. Behav.* **2020**, *215*, 112772. [CrossRef]

21. Di Vincenzo, O.; Marra, M.; Scalfi, L. Bioelectrical impedance phase angle in sport: A systematic review. *J. Int. Soc. Sports Nutr.* **2019**, *16*, 49. [CrossRef]

22. Ballarin, G.; Monfrecola, F.; Alicante, P.; Chierchia, R.; Marra, M.; Sacco, A.; Scalfi, L. Raw Bioelectrical Impedance Analysis Variables (Impedance Ratio and Phase Angle) and Physical Fitness in Cross-Fit® Athletes. In Proceedings of the 8th International Conference on Sport Sciences Research and Technology Support, Budapest, Hungary, 5–6 November 2020; pp. 103–108.

23. Naczk, M.; Kowalewska, A.; Naczk, A. The risk of injuries and physiological benefits of pole dancing. *J. Sports Med. Phys. Fit.* **2020**, *60*, 883–888. [CrossRef] [PubMed]

24. Nicholas, J.C.; McDonald, K.A.; Peeling, P.; Jackson, B.; Dimmock, J.A.; Alderson, J.A.; Donnelly, C.J. Pole Dancing for Fitness: The Physiological and Metabolic Demand of a 60-Minute Class. *J. Strength Cond. Res.* **2019**, *33*, 2704–2710. [CrossRef] [PubMed]

25. Nawrocka, A.; Mynarski, A.; Powerska, A.; Rozpara, M.; Garbaciak, W. Effects of exercise training experience on hand grip strength, body composition and postural stability in fitness pole dancers. *J. Sports Med. Phys. Fit.* **2017**, *57*, 1098–1103. [CrossRef] [PubMed]

26. Fields, J.B.; Metoyer, C.J.; Casey, J.C.; Esco, M.R.; Jagim, A.R.; Jones, M.T. Comparison of Body Composition Variables Across a Large Sample of National Collegiate Athletic Association Women Athletes From 6 Competitive Sports. *J. Strength Cond. Res.* **2018**, *32*, 2452–2457. [CrossRef]

27. D'Alessandro, C.; Morelli, E.; Evangelisti, I.; Galetta, F.; Franzoni, F.; Lazzeri, D.; Piazza, M.; Cupisti, A. Profiling the diet and body composition of subelite adolescent rhythmic gymnasts. *Pediatr. Exerc. Sci.* **2007**, *19*, 215–227. [CrossRef]

28. Jürimäe, J.; Võsoberg, K.; Tamm, A.-L.; Maasalu, K.; Remmel, L.; Tillmann, V. Body composition and inflammatory markers in pubertal girls: Comparison between athletes and non-athletic controls. *Eur. J. Sport Sci.* **2017**, *17*, 867–873. [CrossRef]

29. Marra, M.; Caldara, A.; Montagnese, C.; De Filippo, E.; Pasanisi, F.; Contaldo, F.; Scalfi, L. Bioelectrical impedance phase angle in constitutionally lean females, ballet dancers and patients with anorexia nervosa. *Eur. J. Clin. Nutr.* **2009**, *63*, 905–908. [CrossRef]

30. Barranco-Ruiz, Y.; Ramírez-Vélez, R.; Martínez-Amat, A.; Villa-González, E. Effect of Two Choreographed Fitness Group-Workouts on the Body Composition, Cardiovascular and Metabolic Health of Sedentary Female Workers. *Int. J. Environ. Res. Public Health* **2019**, *16*, 4986. [CrossRef]

31. National Health and Nutrition Examination Survey (NHANES). Anthropometry Procedures Manual. January 2017; pp. 3–6. Available online: https://wwwn.cdc.gov/nchs/data/nhanes/2017-2018/manuals/2017_Anthropometry_Procedures_Manual.pdf (accessed on 15 October 2021).

32. Heymsfield, S.B.; McManus, C.; Smith, J.; Stevens, V.; Nixon, D.W. Anthropometric measurement of muscle mass: Revised equations for calculating bone-free arm muscle area. *Am. J. Clin. Nutr.* **1982**, *36*, 680–690. [CrossRef]

33. Organ, L.W.; Bradham, G.B.; Gore, D.T.; Lozier, S.L. Segmental bioelectrical impedance analysis: Theory and application of a new technique. *J. Appl. Physiol.* **1994**, *77*, 98–112. [CrossRef] [PubMed]

34. Sun, S.S.; Chumlea, W.C.; Heymsfield, S.B.; Lukaski, H.C.; Schoeller, D.; Friedl, K.; Kuczmarski, R.J.; Flegal, K.; Johnson, C.L.; Hubbard, V.S. Development of bioelectrical impedance analysis prediction equations for body composition with the use of a multicomponent model for use in epidemiologic surveys. *Am. J. Clin. Nutr.* **2003**, *77*, 331–340. [CrossRef] [PubMed]

35. Cohen, J. A power primer. *Psychol. Bull.* **1992**, *112*, 155.

MDPI
St. Alban-Anlage 66
4052 Basel
Switzerland
www.mdpi.com

International Journal of Environmental Research and Public Health Editorial Office
E-mail: ijerph@mdpi.com
www.mdpi.com/journal/ijerph